The Effects of Hypergravity and Microgravity on Biomedical Experiments

The Effects of Hypergravity and Microgravity on Biomedical Experiments

Thais Russomano, Gustavo Dalmarco, and Felipe Prehn Falcão

ISBN: 978-3-031-00496-4   paperback
ISBN: 978-3-031-01624-0   ebook

DOI 10.1007/978-3-031-01624-0

A Publication in the Springer series

SYNTHESIS LECTURES ON BIOMEDICAL ENGINEERING #18

Lecture #18

Series Editor: John D. Enderle, University of Connecticut

Series ISSN

ISSN 1930-0328   print
ISSN 1930-0336   electronic

The Effects of Hypergravity and Microgravity on Biomedical Experiments

Thais Russomano, Gustavo Dalmarco, and Felipe Prehn Falcão
Microgravity Centre, Pontifícia Universidade Católica do Rio Grande do Sul

SYNTHESIS LECTURES ON BIOMEDICAL ENGINEERING #18

ABSTRACT

Take one elephant and one man to the top of a tower and simultaneously drop. Which will hit the ground first?

You are a pilot of a jet fighter performing a high-speed loop. Will you pass out during the maneuver?

How can you simulate being an astronaut with your feet still firmly placed on planet Earth?

In the aerospace environment, human, animal, and plant physiology differs significantly from that on Earth, and this book provides reasons for some of these changes. The challenges encountered by pilots in their missions can have implications on the health and safety of not only themselves but others. Knowing the effects of hypergravity on the human body during high-speed flight led to the development of human centrifuges. We also need to better understand the physiological responses of living organisms in space. It is therefore necessary to simulate weightlessness through the use of specially adapted equipment, such as clinostats, tilt tables, and body suspension devices.

Each of these ideas, and more, is addressed in this review of the physical concepts related to space flights, microgravity, and hypergravity simulations. Basic theories, such as Newton's law and Einstein's principle are explained, followed by a look at the biomedical effects of experiments performed in space life sciences institutes, universities, and space agencies.

KEYWORDS

microgravity, weightlessness, hypergravity, clinostats, ground-based simulations, space physiology, centrifuge.

Contents

CHAPTER 1

General Concepts in Physics— Definition of Physical Terms

1.1   GRAVITY AND RELATED CONCEPTS

Konstantin Tsiolkovsky, a pioneering Russian rocket scientist, once said that "The Earth is the cradle of Humanity, but we can not [*sic*] live in a cradle forever." Mankind, in its process of evolution, will eventually leave their home planet searching for new places to inhabit, just like our primitive ancestors did when they left their caves. Living organisms on Earth share at least one thing in common: all of them live under the influence of the gravitational force of the Earth, which produces an acceleration of approximately 9.81 m/s^2 at mean sea level. Earth's acceleration is represented by the small letter "g." Therefore, any variation in this force will result in physiological changes in any given organism. To better understand topics related to the effects of hypergravity and microgravity on living organisms, it is important to be familiar with some definitions of physical terms, theories, and laws that will be presented and discussed in this chapter (Chandler; Cutnell and Johnson, 2006; Dobson et al, 2006; Halliday et al, 1993).

Gravitation (or *gravity*) is the tendency of objects with mass to accelerate toward each other. The terms *gravity* and *gravitation* are often used to explain the same thing, but for many authors there is a definite difference between the two:

Gravitation is the attractive force existing between any two objects that have mass. Gravitation is the reason for the very existence of the Earth, the Sun, and other celestial bodies. Without it, matter would not have coalesced into these bodies and life as we know it would not exist.

Gravitation is also responsible for keeping the Earth and the other planets in their orbits around the sun, the Moon in its orbit around the Earth, the formation of tides, and various other natural phenomena that are commonly observed (Figure 1.1).

Gravity, however, is the gravitation related to Earth. It is then the gravitational force that occurs between the Earth and other bodies, the force acting to pull objects toward the Earth. It is expressed as 1G (capital "G," as opposed to the acceleration *g*). Bodies with less mass than the Earth will have values lower than 1G (the Moon has 1/6G), and bodies with mass bigger than the Earth will have values higher than 1G (planet Jupiter has 3.5G).

FIGURE 1.1: The gravitational force keeps the planets in orbit about the sun (http://en.wikipedia .org/wiki/Gravity).

Mass is a property of a physical object that quantifies the amount of matter and energy it is equivalent to and is expressed by the symbol *m*.

Acceleration (expressed by the symbol *a*) is defined as the rate of change of velocity. It is thus a vector quantity with dimension length/time² (SI units = m/s²), and the instantaneous accelera-tion of an objection is given by Equation 1.1. By being a vector, it must be described with both a direction and a magnitude. It can have positive and negative values—called *acceleration* (increasing speed) and *deceleration* (decreasing speed), respectively, as well as change in direction.

$$a = \frac{\mathrm{d}v}{\mathrm{d}t} \tag{1.1}$$

where *a* is acceleration, *v* is velocity, *t* is time, and *d* is Leibniz's notation for differentiation.

Gravitation is one of the four fundamental interactions in nature, the other three being the electromagnetic force, the weak nuclear force, and the strong nuclear force. Compared to the other three fundamental interactions in nature, gravitation is the weakest one. It, however, acts over great distances and is always present.

Gravitation is interpreted differently by classic mechanics, relativity, and quantum physics. In classic mechanics, gravitation arises out of the force of gravity. In general relativity, it is the mass that curves space time, not a force. In quantum gravity theories, the gravitation is the postulated carrier of the gravitational force, time space itself is envisioned as discrete in nature, or both.

The gravitational attraction of the Earth endows objects with weight that causes them to fall to the ground when dropped (the Earth also moves toward the object but only by an infinitesimal amount).

The Universal Law of Gravitation was postulated by the English physicist and mathematician Sir Isaac Newton (1642–1727). There is a popular story that the origin of this theory happened when Newton was sitting under a tree and an apple fell on his head. This is almost certainly not an exact truth, with embellishment of details, as what happens in most legends. Probably, the more correct version of the story is that Newton, upon observing (or imagining!) an apple fall from a tree, began to think that the apple is accelerated because its velocity changes from zero as it is hanging on the tree then moves toward the ground. He then considered that the same force that pulled the apple toward the Earth is the same one that makes the Moon to orbit our planet.

Newton's theory of how a celestial body can orbit another celestial body can be illustrated by his well-known example shown in Figure 1.2. Suppose that a cannon ball is fired horizontally from a high mountain on top of the Earth. The projectile will eventually fall to the ground, as indicated by the shortest trajectory in the figure. As the velocity is increased for projectile fired by the imaginary cannon, it will travel further and further before returning to Earth. Finally, Newton reasoned that if the cannon projected the cannon ball with exactly the right velocity, the projectile would travel completely around the Earth, always falling in the gravitational field but never reaching the

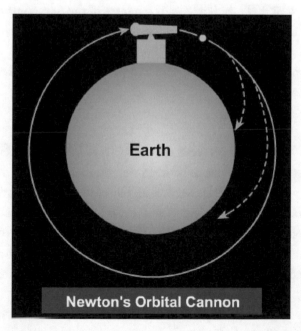

FIGURE 1.2: Newton's imaginary cannon experiment (http://csep10.phys.utk.edu/astr161/lect/history/newtongrav.html).

Earth, which is curving away at the same rate that the projectile falls. That is, the cannon ball would have been put into orbit around the Earth. This was the basis for the theory regarding the orbit of the Moon around the Earth. Newton concluded that the Moon continuously "fell" in its path around the Earth because of the acceleration due to gravity, thus producing its orbit. Newton then concluded that any two objects in the Universe exert gravitational attraction on each other, with the force having a universal form.

He published the Universal Law of Gravitation in the *Philosophiae Naturalis Principia Mathematica* (1687) (Latin for "mathematical principles of natural philosophy"), which also contained Newton's laws of motion that is the foundation of classic mechanics.

The Universal Law of Gravitation of Newton states that, "every particle in the universe attracts every other particle with a force that is directly proportional to the product of their masses and inversely proportional to the square of the distance between them." As expressed in Equation 1.2, if the particles have masses m_1 and m_2 and are separated by a distance r (from their centers of gravity), the magnitude of this gravitational force is:

$$F = G \, \frac{m_1 m_2}{r_2} \tag{1.2}$$

where F is the magnitude of the gravitational force between the two point masses, G is the gravitational constant (6.67×10^{-11} N $\cdot$ m^2/kg^2), m_1 is the mass of the first point mass, m_2 is the mass of the second point mass, and r is the distance between the two point masses.

The Austrian–Czech physicist and philosopher Ernest Mach and the German physicist Albert Einstein have concluded that gravitation and acceleration are equivalent and produce the same effects on a body of any nature.

1.2 WEIGHT VERSUS MASS AND INERTIAL MASS

It is essential to understand the difference between weight, mass, and inertial mass. *Mass* per se is the measure of inertia, whereas *weight* is the force an object feels due to gravitational attraction.

The mass of an object is independent of where it is located. Weight is very much dependent on location. For example, near the surface of the Earth, the weight of an object is $W = mg$, where $g = 9.81$ m/s^2 directed toward the center of the Earth. However, near the surface of the Moon, the weight will be calculated in the same way, but the value of g is 1.6 m/s^2 directed toward the center of the Moon.

Inertial mass is mainly defined by Newton's law ($F = ma$), which states that when a force F is applied to an object, it will accelerate proportionally, and that constant of proportion is the mass of that object. Therefore, to determine the inertial mass, it is necessary to apply a force of F (N) to

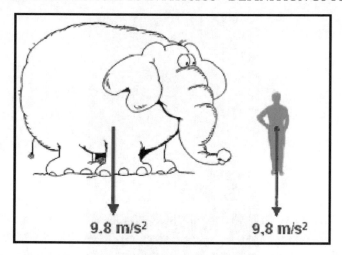

FIGURE 1.3: Mass and inertial mass causing the same acceleration (http://exploration.grc.nasa.gov/ Exploration/redirect.htm).

an object, to measure the acceleration a (m/s^2), and to divide F/a, which gives the inertial mass m (kg). It is easy to conclude that the bigger the mass of a body, the bigger will be its inertia, which is a unique property of gravity.

A classic example is presented here. Imagine an elephant with a mass of 700 kg and a person with a mass of 70 kg. By using the formula $F = ma$, it is possible to calculate the force of attraction between the Earth and the elephant ($F = 700 \times 9.8 = 6860$ N) and between the Earth and the person ($F = 70 \times 9.8 = 686$ N), as represented in Figure 1.3. However, the g force produced in both cases will be the same due to the inertial mass of each one, the elephant and the man. If F is 686 N and the mass of the man is 70 kg, a is 9.8 m/s^2. Equation 1.3 expresses the calculation of a for the elephant.

$$F = m \cdot a \quad a = \frac{F}{m} \quad a = \frac{6860}{700} \quad a = 9.8 \text{ m/s}^2 \tag{1.3}$$

1.3 APPARENT WEIGHT AND NORMAL FORCE

On Earth, man is aware of his weight because there is an external medium (the Earth's surface) pushing back against his body with an inertial force equal to the gravitational force attracting it toward the center of the Earth.

According to Newton, whenever objects A and B interact with each other, they exert forces upon each other. When you stand on the floor, your body exerts a downward force on the floor

FIGURE 1.4: Newton's third law of motion: action and reaction forces (http://csep10.phys.utk.edu/astr161/lect/history/newtongrav.html).

(*weight*) and the floor exerts an upward force on your body (*normal force*), as illustrated in Figure 1.4. These two forces are called *action* and *reaction forces* and are the subject of Newton's third law of motion. Formally stated, Newton's third law is: "For every action, there is an equal and opposite reaction."

The normal force produces the feeling of an apparent weight, which is perceived by the musculoskeletal system and transmitted to the central nervous system. Imagine then a person standing on a weighing machine placed inside an elevator (Figure 1.5).

The weight or, as better called, the *apparent weight* of the person will be different in each situation presented in Figure 1.5. When the elevator is at rest, weight and normal forces are equal (Figure 1.5a) because the normal force on the person's feet balances the person's weight. If the elevator is accelerating upward (Figure 1.5b) or downward with $a < g$ (Figure 1.5c), the normal force is greater than the weight (*the person's apparent weight increases, making the person feel heavier*) and less than the weight (*the person's apparent weight decreases, making the person feel lighter*), respectively. If, however, the elevator is accelerating downward with $a = g$ (Figure 1.5d), the normal force and apparent weight will be zero. On the other hand, when the elevator is accelerating downward with $a > g$ (Figure 1.5e), the person will be thrown against the elevator's ceiling and the normal force will be there and not on the floor of the elevator anymore. In this case, the normal force will have a negative module.

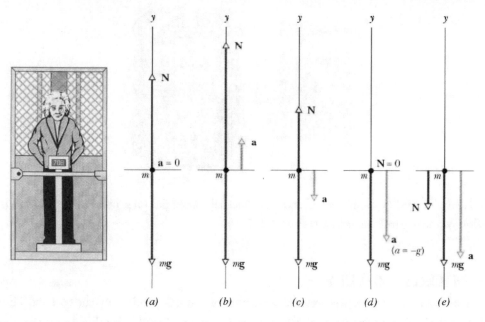

FIGURE 1.5: A person standing on a weighing scale placed inside an elevator: at rest (a), moving up (b), moving down with $a < g$ (c), moving down with $a = g$ (d), and moving down with $a > g$ (e) (http://exploration.grc.nasa.gov/Exploration/redirect.htm).

1.4   THE EINSTEIN PRINCIPLE

The equivalence principle has historically played an important role in the development of gravitation theory. The origins of the equivalence principle began with Galileo Galilei demonstrating in the late 16th century that objects accelerated toward the center of the Earth at the same rate. Newton codified this with his gravitational theory in which it was postulated that inertial and gravitational masses are the same. The equivalence principle proper was introduced by Albert Einstein (1879–1955) in 1907. At that time, he made the observation that the acceleration of bodies toward the center of the Earth with acceleration of $1g$ ($g = 9.81$ m/s²) is equivalent to the acceleration of inertially moving bodies that one would observe if one was on a rocket in free space accelerating at a rate of $1g$. Einstein thus stated: "we [. . .] assume the complete physical equivalence of a gravitational field and a corresponding acceleration of the reference system."

Therefore, according to Einstein, the gravitational "force" as experienced locally while a person, for example, is standing on a massive body (such as the Earth) (Figure 1.6a) is actually the same as the *pseudoforce* experienced by an observer in a noninertial (accelerated) frame of reference (Figure 1.6b). In both cases, the person is not able to identify the situation because he is under the same acceleration independently of its gravitational or inertial origin.

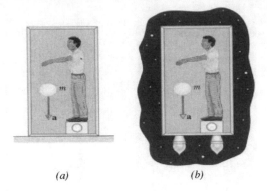

(a)   *(b)*

FIGURE 1.6: The effect of an inertial and a noninertial referential on a person is equivalent (http://exploration.grc.nasa.gov/Exploration/redirect.htm).

1.5   MICROGRAVITY

The term *microgravity* can be interpreted in a number of ways, depending on the context. The prefix micro (μ) is derived from the original Greek *mikros*, meaning "small." By this definition, a microgravity environment is one that imparts to an object a net acceleration that is small compared with that produced by Earth at its surface, which can be achieved by using various methods, including Earth-based drop towers, parabolic aircraft flights, and Earth-orbiting laboratories. In practice, such accelerations will range from approximately 1% of Earth's gravitational acceleration (aboard aircraft in parabolic flight) to better than one part in a million (e.g., aboard Earth-orbiting free flyers).

In general terms, there are three ways for microgravity to occur:

1. To have a celestial body in a *hypothetical place in the Universe* so far away from any other celestial body that the gravitational force would be null.

2. To have a celestial body placed in a strategic point between two celestial bodies, called *Lagrangian points*. Joseph-Louis Lagrange (born Guiseppi-Luigi Lagrangia in Turin, Italy, in 1736) was a brilliant French mathematician and astronomer. Typical of his work was his thorough analysis, in 1772, of the "three-body problem." It deals with the forces acting on a small mass located near two other masses, one larger than the other. The equations he developed are known as "the Lagrangians." They have become fundamental to the design of all spaceflight trajectories. His mathematical solutions to this problem revealed that there are five stable points in the neighborhood of the two large bodies in a typical sun–planet combination. These are the points at which the gravitational accelerations balance each other and where small objects would tend to remain in place. Lagrange referred to them as "Libration points" and labeled them L1 to L5.

With the advent of the Space Age, Lagrange's equations and these five points have become of enormous importance. L1 to L5 are now called the *Lagrangian points*. L1, L2, and L3 are on the straight line passing through the Sun and the planet. L1 lies on this line at the point of equal gravitational attraction by the Sun and by the planet and is therefore closer to the planet. L2 is on the same line, extended, at a point an equal distance beyond the planet. L3 is behind the Sun and lies slightly outside the orbit path of the planet. L4 and L5 lie on the planet's orbit path around the Sun. Lines from them to the sun form 60° angles with the sun–planet axis.

Consider, for example, Earth's Lagrangian points:

- L1 is 1.5 million kilometers from Earth on the approximately 150-million-kilometer line joining Earth to the sun. L1 is being used as a parking spot for spacecraft whose purpose is to monitor the Sun. The US/European spacecraft "SOHO" (Solar and Heliospheric Observatory) is an example.
- L2 is also situated 1.5 million kilometers from Earth and it is permanently in the shadow of the Earth. L2 is about four times more distant from us than the Moon. It can be used for deep-space observatories and for communications stations.
- L3, located behind the Sun, is blocked by it and has no foreseeable value.
- L4 and L5, however, are destined to be used as parking spots for spacecraft involved in lunar and planetary expeditions in the Solar System because a spacecraft on L4 and L5 would tend to remain in place.

3. Microgravity can also be achieved with a number of technologies, each depending on the act of free fall, such as drop towers or drop tubes, parabolic flights, small rockets, spacecrafts, and space stations.

However, it is common to think that there is no gravity above the Earth's atmosphere and this is why there appears to be no gravity aboard orbiting spacecraft. Typical orbital altitudes for human spaceflight range from 120 and 360 miles (from 192 to 576 km) above the surface of the Earth. The gravitational field, however, is still strong in these regions because this is only approximately 1.8% the distance to the Moon. The Earth's gravitational field at approximately 250 miles (400 km) above the surface maintains 88.8% of its strength at the surface. Therefore, orbiting spacecrafts, like the Space Shuttle or Space Station, are kept in orbit around the Earth by gravity.

Once a spacecraft is in orbit above the effective limits of the atmosphere, it is in a state of free fall about the Earth because of the continuous accelerative force of gravity pulling it toward the center of the planet. The spacecraft does not fall back to Earth due to its orbital velocity (approximately 27,000 km/h). The velocity produces tangential and inertial forces, which are needed to counterbalance the force of gravity. When this state of equilibrium is attained, the spacecraft and the astronauts are in microgravity. For spaceflights to the Moon or interplanetary spaceflights, the velocity required to escape Earth's gravity is 40,000 km/h (escape velocity).

The term *microgravity* is preferred to the more familiar weightlessness. As said before, even in space, conditions of zero gravity do not exist in spaceflight and the astronauts are normally exposed to a microgravity of 1×10^{-4} to 1×10^{-5} G, which increases to 1×10^{-3} G during spacecraft maneuvers. However, the terms *microgravity* and *weightlessness* will be used exchangeable in this book.

The Zero Gravity Research Facility (Zero-G) is NASA's premier facility for conducting ground-based microgravity research and is the largest one of its kind within the United States. Operational since 1966, it is one of two drop towers located at NASA Glenn Research Center. Built during the "Space Race" era of the 1960s, it was originally designed to support research and development involving spaceflight components and fluid systems in a weightless environment.

The drop tower of Bremen, Germany (Figure 1.7), features a 110-m-high drop chamber with a diameter of 3.5 m that can be evacuated. An ensemble of 18 pumps allows the attainment of a residual pressure in the chamber of 1 Pa within 2.5 h. The chamber is free-standing inside a concrete building, which protects it from wind perturbations. The experiments are done in a pressurized capsule, 81 cm in diameter and firmly attached to its structure. After integration, the capsule is elevated to the release platform where it is held with an electropneumatic suspension device during the evacuation of the drop chamber. The drop is initiated by opening the suspension device, which generates very limited perturbations to the capsule. A free-fall duration of 4.7 s with residual accelerations lower than $10^{-5}g$ is thus achieved. Deceleration occurs in an 8-m-deep tank filled with polystyrene pellets. The deceleration levels do not exceed 50g. In general, it is possible to perform two to three drops a day when an experiment is being conducted.

In general terms, parabolic flights are used to conduct short-term microgravity scientific and technological investigations, to test instrumentation before use in space, to validate operational and experimental procedures, and to train astronauts for future space missions.

NASA started its parabolic flight program, nicknamed the *Vomit Comet Program*, in 1973. Twin KC-135 Stratotankers were used until December 2004 and have since been retired. A converted Boeing 707, known as NASA 930, was also used by Universal Pictures and Imagine Entertainment for filming scenes involving weightlessness in the movie *Apollo 13*. This aircraft retired in 2000. In 2005, NASA replaced the KC-135 with a McDonnell Douglas C-9 that was formerly owned by KLM Airlines and the US Navy. In the same year, the Zero-Gravity Corporation, a commercial parabolic flight operator that offers parabolic flight to both researchers and adventure tourists, began flying parabolic flights for NASA with Boeing 727 jets.

A typical NASA parabolic flight lasts 3 h and carries experiments and crewmembers to a beginning altitude approximately 7 km above sea level. The plane climbs rapidly at 45° (pull up),

FIGURE 1.7: The drop tower in Bremen, Germany (http://www.spaceflight.esa.int).

traces a parabola (pushover), and then descends at a 45° (pull out). During the pull-up and pull-out segments, crew and experiments experience between $2g$ and $2.5g$. During the parabola, at altitudes ranging from 7.3 to 10.4 km, net acceleration drops as low as $1.5 \times 10^{-2}g$ for more than 15 s.

The European Space Agency (ESA) began flying parabolic flight campaigns in 1984. From this time up to 1988, a total of six campaigns were carried out from the Ellington airfield in Houston (TX) using a NASA KC-135 aircraft. In 1988, the French space agency CNES made its Caravelle zero-gravity aircraft available to ESA, and between this time and 1995, 15 ESA parabolic campaigns were conducted. Once, in 1994, ESA also flew one campaign on a Russian Ilyushin IL-76 MDK. In 1996, ESA performed its 23rd campaign out of Bordeaux using a NASA KC-135 aircraft and in September 1997 performed its 24th campaign using the newly available Airbus A-300 Zero-G aircraft of CNES, and it is this aircraft that has been used ever since. The Airbus A-300 Zero-G is operated out of the Bordeaux-Mérignac airport by the company Novespace. ESA parabolic flight profile is practically the same as the one used by NASA, differing in the level of hypergravity and the altitude during pull up and pull out, as shown in Figure 1.8.

In May 2006, the German Aerospace Centre (DLR) carried out its first parabolic flight with the Novespace's Airbus A300 ZERO-G from Cologne-Bonn International Airport. The current parabolic flight program accessible to Canadian experimenters is centered on the use of the

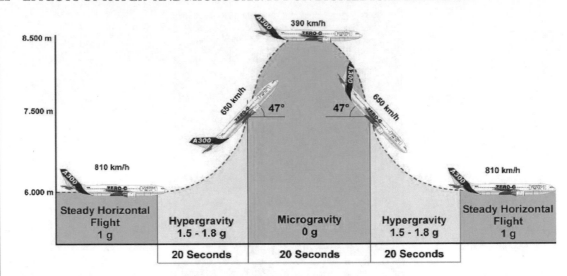

FIGURE 1.8: ESA parabolic flight profile (http://www.esa.int).

National Research Council's (NRC) Falcon 20 aircraft. The Flight Research Laboratory of the NRC implemented the first Falcon 20 parabolic flight program in 1991. The first microgravity experiment onboard a parabolic aircraft flew in December 1993. The Canadian Space Agency continued to use the Falcon 20 until 1995 when, through an agreement with NASA, gained access to their DC-9 parabolic aircraft. The Falcon 20 is a commercial jet with modified hydraulic and fuel systems to allow it to perform parabolic maneuvers. The parabolas are executed in a restricted area, approximately 4000 and 8000 m in altitude. A typical flight currently consists of four parabolic trajectories that have a total duration of approximately 45 min. Each parabola lasts approximately 75 s, of which 15 to 20 s are at $0.02g$ or less, followed by a $1.8g$ pull-out. Other aircrafts also available to Canadian scientists for microgravity experiments are NASA's KC-135 and Novespace's Airbus A300.

A sounding rocket follows a suborbital trajectory and can produce several minutes of free fall. The period of free fall exists during its coast, after burn out, and before entering the atmosphere. Acceleration levels are usually at or below $10^{-5}g$. NASA has used many different sounding rockets for microgravity experiments. The most comprehensive series of launches used SPAR (Space Processing Application Rocket) for fluid physics, capillarity, liquid diffusion, and electrolysis experiments from 1975 to 1981. The SPAR could lift 300-kg payloads into freefall parabolic trajectories lasting 4 to 6 min.

Because this is a cheaper way to create microgravity on Earth, this is the choice for developing countries, such as Brazil. On May 15, 1999, for example, three Brazilian University experiments along with a variety of other scientific and commercial microgravity experiments flew onboard the Brazilian VS-30 rocket, in a mission named *Operation São Marcos*, organized by the Instituto de Aeronáutica e Espaço, Centro Tecnico Aeroespacial and launched from the Alcântara Launch Center. The Brazilian Instituto Tecnológico de Aeronáutica flew two of its microgravity devices—Materials Dispersion Apparatus (MDA). The MDAs are brick-sized mini-laboratories to conduct a wide variety of science experiments in the microgravity environment. One MDA was dedicated to Brazilian University microgravity experiments as part of an ongoing space education program. The Brazilian experiments include the effect of launch vibration on special lipase emulsions for future space missions, the effect of low gravity on *Planaria* worms, and the crystallization of biomedical products for future antibiotic drugs (http://www.inpe.br).

1.6 HYPOGRAVITY

The acceleration due to gravity at the surface of a planet varies directly as the mass and inversely as the square of the radius. This follows directly from Newton's law of universal gravitation.

The Moon is 384,403 km away from the Earth. Its diameter is 3476 km. The acceleration due to gravity is 1.62 m/s² because the Moon has less mass than Earth. It is approximately one sixth that of the acceleration due to gravity on Earth, that is, 9.81 m/s².

Mars and Earth have diameters of 6775 and 12,775 km, respectively. The mass of Mars is 0.107 times that of Earth. This makes the gravitational acceleration on Mars (gm) 3.73 m/s², as expressed in Equation 1.4:

$$gm = 9.81 \times 0.107 \times (12{,}775/6775)^2 = 3.73 \text{ m/s}^2 \qquad (1.4)$$

Therefore, if a body weighs 200 N on Earth, it is possible to calculate how it would weigh on Mars. Knowing that the weight of an object is its mass (m) times the acceleration of gravity, we can have: $W = m \times g$; $200 = 9.81 \times m$; $m = 20.41$ kg. This mass is the same on Mars, so the weight on Mars will be 3.73×20.41, which is equal to 76.1 N.

The change in body weight can affect the human posture. Figure 1.9 illustrates the change in body posture of an individual on Earth, on Mars and in space (microgravity).

1.7 PARTIAL-GRAVITY ENVIRONMENTS ON EARTH

The three primary techniques to simulate partial gravity are the underwater immersion, parabolic flights, and body suspension devices.

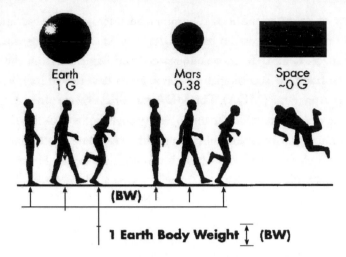

FIGURE 1.9: Schematic representation of body posture in different gravitational environments (http://www.nasa.gov).

1.7.1   Underwater Immersion

During tests, a neutrally buoyant subject is ballasted to simulate the desired partial gravity loading. For example, one sixth of the subject's body mass is added in ballast if a lunar simulation is desired. Water immersion offers the subject freedom from time constraints and freedom of movement, but the hydrodynamic drag is disadvantageous for several physiological studies, especially the ones that include body movements.

1.7.2   Parabolic Flights

The NASA KC-135 aircraft or Russian IL-76 aircraft is typically used to simulate partial gravity by flying Keplerian trajectories through the sky. This technique provides approximately 30 and 40 s for lunar gravity and martian gravity tests, respectively. Parabolic flight is the only way to effect true partial gravity on Earth, but experiments are expensive and of limited duration.

1.7.3   Body Suspension Devices

The cable-suspension method typically uses vertical cables to suspend the major segments of the body and relieve some of the weight exerted by the subject on the ground, thus simulating partial gravity. Suspension systems often afford the most economical partial-gravity simulation technique but limit freedom of movement. Many partial-gravity suspension systems have been designed and used since the Apollo program, for example.

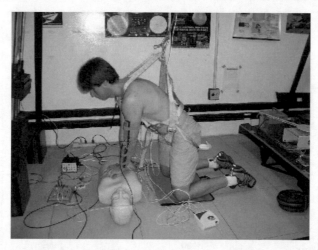

FIGURE 1.10: Male subject performing external chest compression while wearing the harness for hypogravity simulation using the Microgravity Centre/KCL BSD system (http://www.pucrs.br/feng/microg).

The Microgravity Centre/FENG-PUCRS, Brazil, and the Aerospace Medicine Group of King's College London, UK, have developed a body suspension device (BSD) system to study cardiopulmonary resuscitation during microgravity and hypogravity (lunar and martian) simulations. The BSD system is composed of a body harness, counterweights, and a load cell. The structure is pyramidal and consists of steel bars with a thickness of 6×3 cm. It has a rectangular base area of 300×226 cm and a height of 200 cm. A steel cable connects the counterweights through a system of pulleys to a harness worn by the subject. It uses a counterweight system made of 20 bars of 5 kg each, placed opposite the subject (Figure 1.10). Results of the experiments during Mars and Moon simulation will be described later.

1.8   HYPERGRAVITY

Newton's law states that a body will retain its velocity unless a net force acts on it. Thus, if a body is travelling in a circle, a force must be acting on it to prevent it from travelling in a straight line. This force is the centripetal force (F_c), the only force necessary for a circular motion. What is interpreted sometimes as a centrifugal force is the tendency of the object to follow a straight line, which would bring it outside of its circular trajectory. It is equal in magnitude but opposite to the centripetal force required to constrain the body to move in a circular motion. Physicists, however, commonly do not recommend the use of the term *centrifugal force* to prevent people drawing wrong conclusions from the idea of the experience of a centrifugal force. For practical reasons, the term *centrifuge* will be used here because it still is in the areas of aerospace physiology and medicine nowadays.

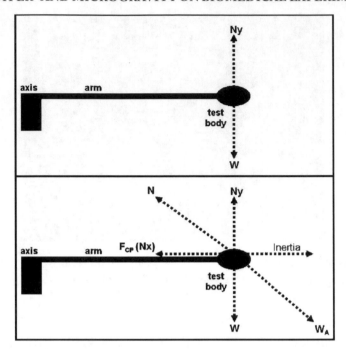

FIGURE 1.11: Schematic view of the physical forces acting on a test body at rest (above) and during rotation (bottom) (http://www.pucrs.br/feng/microg).

Based on what was explained before, the weight (W) acting on a test body at rest is counteracted by the normal force (N), which avoids the body falling down. During rotation, however, a centripetal force (F_{CP}) starts acting on the test body pulling it toward the center of the circle, whereas the inertia keeps the test body in a tangential direction. The resultant force, that is, the apparent weight (W_A) produced, is related to the G effect applied on the test body, as illustrated in Figure 1.11 and expressed in Equations 1.5 and 1.6.

$$W_A{}^2 = F_{CP}{}^2 + W^2 \qquad (1.5)$$

$$G = W_A / W \qquad (1.6)$$

· · · · ·

CHAPTER 2

The Effects of Hypergravity on Biomedical Experiments

2.1   HYPERGRAVITY AND HUMAN PHYSIOLOGY

The first reported disturbance due to G's occurred in 1918, when the pilot of a Sopwith Triplane noted that the sky appeared to be gray (gray-out) just before he fainted during a tight turn at +4.5Gz. In 1927, Jimmy Doolittle, as part of his graduate studies in aeronautical engineering at the Massachusetts Institute of Technology, mounted a recording accelerometer in a Fokker PW aircraft and logged G's during aerobatic maneuvers. He reported that in a sustained +4.7Gz condition (power spiral), he began to lose his sight and, for a short time, everything went black (blackout). He retained all faculties except sight and had no difficulty in righting the airplane. With the growing use of military aircrafts, there was an increased interest in the effects of G's on pilots. A need to study the effects of acceleration under controlled conditions led to the development of large centrifuges in which human subjects could be exposed to G's while they attempted to perform certain flying tasks. The development of higher performance aircraft encouraged the development of strategies for combating the effects of G forces. Nowadays, anti-G suits and muscle tensing or grunting procedures (M-1, L-1 maneuvers), among other things, are used to protect pilots and crews from effects of increased G force on human physiology.

An axial nomenclature system has been used to better explain the concept of human body G vectors and is the basis for studies related to acceleration physiology (Gell, 1961). The three major axes are longitudinal (Z), lateral (Y), and horizontal (X). The direction of acceleration forces along the axes is called (+) or (−). The inertial forces are opposite to the acceleration forces, as indicated in Figure 2.1.

The G effect on human physiology occurs in speeding up (acceleration) or slowing down (deceleration), such as a pull-up from a dive. The amount of G experienced by the pilot depends on how vigorously the pilot pulls back on the controls and how readily the aircraft responds. Assume, however, that the pilot "pulls" +4Gz; if the pilot were on scales, he or she would appear to weigh four times his or her usual weight. Now, imagine a pushover at the start of a dive. The aircraft changes

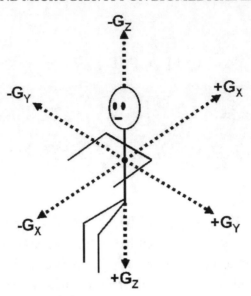

FIGURE 2.1: Standard acceleration nomenclature. The arrows indicate the direction of the inertial reaction to an equal and opposite acceleration (Ernsting et al, 1999).

direction in such a way that the pilot tends to be thrown upward and outward. The pilot may have the sensation of weightlessness.

Symptoms related to G effect on human physiology will depend on duration, direction, area, magnitude of exposure, and individual characteristics. They are more pronounced on the Z axis and are summarized in the following sections.

2.1.1   Positive Gz Effects

The +Gz induced effects may be described as follows:

1. Gray-out. There is graying of vision caused by diminished flow of blood to the eyes. Although there is no associated physical impairment, this condition should serve as a warning of a significant impairment of blood flow to the head. It happens, in general, when a person is subjected from +3 to +4Gz.

2. Blackout. Vision is completely lost. This condition results when the oxygen supply to the light-sensitive retinal cells is severely reduced. Contrary to other common usages of the term, consciousness is maintained. In blackout, some mental activity and muscle function remains, thus the occurrence of blackout warns of seriously reduced blood flow to the head and of a high risk of loss of consciousness. Hearing and mental orientation are

adequate for communication until consciousness is finally lost, leading the individual to the next condition, the loss of consciousness.

3. Loss of consciousness (G-LOC). When the blood flow through the brain is reduced to a certain level, the pilot will lose consciousness, having jerking and convulsive movements. The pilot will slump in his or her seat. Possibly, the pilot will fall against the controls, causing the aircraft to enter flight configurations from which it cannot recover even if consciousness is regained. In centrifuge studies, many pilots lost (and regained) consciousness without realizing they had done so. In an unprotected relaxed subject, unconsciousness commonly occurs between +5Gz and +6Gz.

The most important factor for these effects on vision and consciousness is related to the cardiovascular system. A mean arterial pressure of 100 mmHg at the level of the heart is decreased to 75 mmHg at the level of the brain. On the other hand, pressures below the level of the heart are increased by the hydrostatic pressure and can reach 200 mmHg. At +4Gz, the mean arterial blood pressure at the brain level is expected to be zero because the hydrostatic pressure opposing flow will

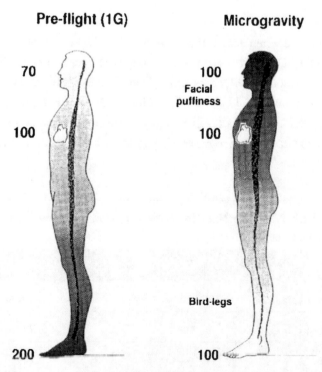

FIGURE 2.2: Schematic view of arterial blood pressure values on Earth (preflight) and during microgravity exposure (http://www.pucrs.br/feng/microg).

be increased fourfold to 100 mmHg. In microgravity, during a space mission, the arterial pressure will be the same from the brain to the feet. Figure 2.2 shows a schematic view of arterial blood pressure distribution during preflight and microgravity.

In a series of studies of pilots in centrifuges, the pilots were unconscious for an average of 15 s. After this, there was an additional 5- to 15-s interval of disorientation. Thus, if there is loss of consciousness due to +Gz forces, there will be a 20- to 30-s (or longer) period during which the pilot is not in control of his or her aircraft.

Because of the number of factors involved, it is difficult to predict how much acceleration a certain individual can withstand. Tolerance is related to the rate of onset of acceleration and to the duration of exposure. Individual tolerance depends on factors, such as the height of the person, age, elasticity of the blood vessels, training, the responses of the heart and blood vessels, and on health. Therefore, because of the variables involved, the centrifuge data in the following table are useful only as an estimate of the average civilian pilot's tolerance to +Gz. These data were collected from 1000 Naval aviation pilots and aviation personnel and apply to rates of onset of about +1G/s—a rate that well may be encountered in civil aerobatic maneuvers (Table 2.1).

2.1.2 Negative Gz Effects

Negative Gz is encountered when acceleration is in a foot to head direction, such as might be obtained during inverted flight, or during an outside loop or pushover maneuver. Blood is then pushed toward the head, and the amount of blood returning from the head is diminished, so the blood tends to stagnate, particularly in the head. Under mild conditions of −Gz forces, the pilot will feel congestion, as when standing on his or her head. The engorgement of blood vessels can cause a reddening or flushing of the facial skin. During negative Gz exposure, blood vessels in the eyes will become

TABLE 2.1: Thresholds in relation to +Gz tolerance (http://www.faa.gov)

Symptom	AVERAGE		
	Threshold	SD	Range
Gray-out	4.1G	±0.7G	2.2–7.1G
Blackout	4.7G	±0.8G	2.7–7.8G
Unconsciousness	5.4G	±0.9G	3.0–8.4G

dilated and a condition termed *redout* may occur. This may be due in part to congestion but may also happen when the lower eyelid, reacting to −Gz, rises to cover the pupil, so that one sees light through the eyelid. Some persons may experience a headache.

Little is known about the effects of high −Gz on humans because −Gz accelerations have caused considerable discomfort in those studied. Aerobatic pilots have reported small hemorrhages in the eyes and skin.

The blood vessels in the brain tolerate mild −Gz stresses well, but the increased blood pressure in the chest and neck causes a slowing of the heart in virtually all subjects. The slowing of the heart and irregularities of beats can add to the stagnation of blood in the brain. Thus, it appears that the greatest threat from −Gz is the loss of consciousness from the slowing of the heart, irregularities of the heartbeats, and stagnation of blood in the head.

2.2   G IMPACT IN SPACE MISSIONS

In the Mercury and Gemini programs, astronauts were exposed to +6Gz to +8Gz during launch and reentry. Launch acceleration loads were around +5Gz in the Apollo program (Figure 2.3), and crewmembers routinely reported that the launches produced no adverse physiological stresses. Maximum reentry +Gz levels for all Apollo missions are shown in Table 2.2.

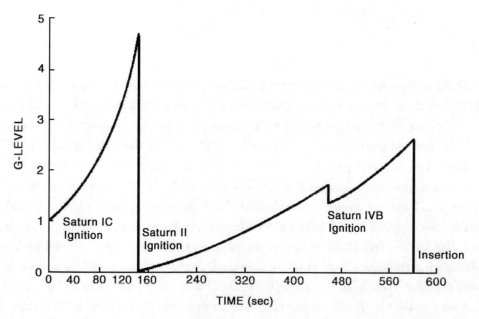

FIGURE 2.3: Typical Apollo launch profile—Saturn V launch vehicle (http://www.nasa.gov).

TABLE 2.2: Apollo-manned spaceflight reentry +Gz levels (http://www.nasa.gov)

FLIGHT	MAXIMUM +GZ AT REENTRY
Apollo 7	3.33
Apollo 8	6.84
Apollo 9	3.35
Apollo 10	6.78
Apollo 11	6.56
Apollo 12	6.57
Apollo 13	5.56
Apollo 14	6.76
Apollo 15	6.23
Apollo 16	7.19
Apollo 17	6.49

The Apollo spacecraft landing system used three parachutes and the repositioned Command Module system used in the Gemini Program. The spacecraft entered the water at a 27.5° angle on a nominal landing. The most severe impact experienced in an Apollo spaceflight occurred with Apollo 12. It was estimated that the Command Module entered the water at a 20° to 22° angle, which resulted in a +15Gz impact, due to the wind that caused the spacecraft to swing and meet the wave slope at the more normal angle. The +15Gz impact of Apollo 12 was described as "very hard" by the crewmen. However, no significant physical difficulties were experienced. Apollo landing impact studies involving 288 human tests were conducted on a linear decelerating device at Holloman Air Force Base. These tests involved impact forces up to +30Gz at various selected body orientations. Although significant effects to the neurological, cardiorespiratory, and musculoskeletal systems were recorded, none of the tests resulted in significant incapacitation or undue pain. Normal Apollo reentry profiles on Earth and on the Moon are shown in Figures 2.4 and 2.5, respectively.

In the Space Shuttle program (Figure 2.6), shortly after clearing the tower, the spacecraft begins a roll-and-pitch program so that the vehicle is below the external tank and solid rocket

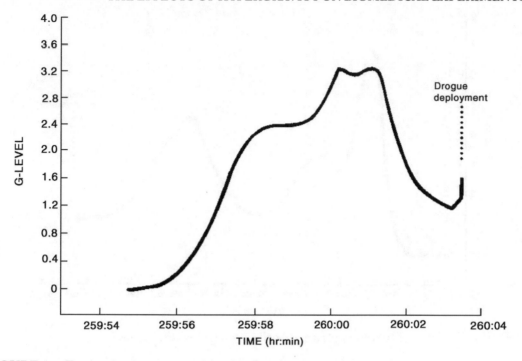

FIGURE 2.4: Earth orbital reentry profile—Apollo 7 (http://www.nasa.gov).

boosters (SRBs). The vehicle climbs in a progressively flattening arc, accelerating as the weight of the SRBs and main tank decrease. Orbital velocity at the 380-km (236-mile) altitude of the International Space Station is 7.68 km/s (27.648 km/h, 17.180 mph), equivalent to Mach 23. For missions toward the International Space Station, the shuttle must reach an azimuth of 51.6° inclination to rendezvous with the station.

Approximately 126 s after launch, the SRBs are released and small separation rockets push them laterally away from the vehicle. The SRBs parachute back to the ocean to be reused. The Shuttle then begins accelerating to orbit on the Space Shuttle main engines. Finally, in the last tens of seconds of the main engine burn, the mass of the vehicle is low enough that the engines must be throttled back to limit vehicle acceleration to +3Gx.

The vehicle begins reentry by firing the Orbiter Manoeuvering System engines in the opposite direction to orbital motion for approximately 3 min. The entire reentry, except for lowering the landing gear and deploying the air data probes, is then under computer control. However, the reentry can be and has (once) been flown manually.

The vehicle starts significantly entering the atmosphere at approximately 400,000 ft. (120 km) at around Mach 25 (8.2 km/s). When the approach and landing phase begins, the Orbiter is at 10,000 ft. (3048 m) altitude, 7.5 miles (12.1 km) to the runway. The pilots apply aerodynamic

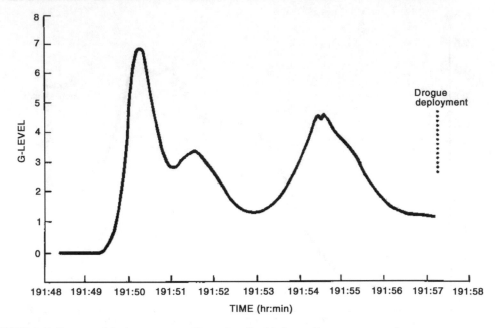

FIGURE 2.5: Lunar orbital reentry profile—Apollo 10 (http://www.nasa.gov).

braking to help slow down the vehicle. The Orbiter's speed is reduced from 424 mph (682 km/h) to approximately 215 mph (346 km/h) (compared to 160 mph for a jet airliner) at touchdown. The landing gear is deployed while the Orbiter is flying at 267 mph (430 km/h). To assist the speed brakes, a 4-ft. (12.2-m) drag chute is deployed once the nose gear touches down at approximately 213 mph (343 km/h). It is jettisoned as the Orbiter slows through 69 mph (111 km/h). During reentry, astronauts are subjected to approximately +3Gz.

2.3   HUMAN CENTRIFUGES

Centrifuges have been widely used around the world to create hypergravity situations, where acceleration in greater than the one on Earth. Human centrifuges are mainly used for aircrew training, physiological and medical research, and equipment testing and evaluation. A future application will be to counteract the effects of microgravity on the human body during long-term space missions.

Erasmus Darwin (1731–1802), the grandfather of Charles Darwin, published in 1795 a work entitled *Zoonomia*, related to sleep and its disorders. In this book, he mentioned a report from Mr. Brindley in which he describes a man who slept deeply when rotated in a machine used to smash corn. Erasmus then concluded that the blood was pooled in the lower limbs leading to unconsciousness. He also believed that centrifugation could be used to treat some medical conditions, such as sleep disorders, heart problems, mental diseases, and fever (Figure 2.7).

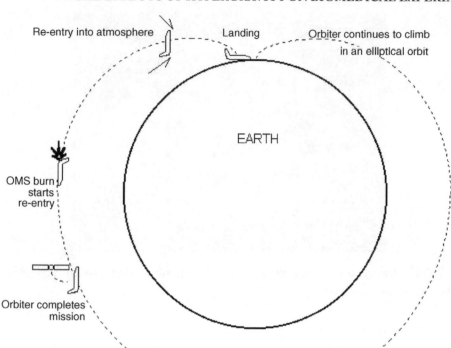

FIGURE 2.6: Schematic summary of Space Shuttle launch and landing profile (http://www.nasa.gov).

In 1903, Sir Hiram Maxim built a device for an amusement park able to reach +6.7Gz. The engineer of the project was unconscious during a test, and it was considered the first reported case of G-LOC. During World War I, fighter planes were used for combat and pilots were exposed to +Gz. Human centrifuges were then necessary to train these fighter pilots. The first one in the United States was in operation in 1935. HG Armstrong and JW Heim conducted a series of experiments to study human physiology under +Gz, publishing the results in the *Effects of Acceleration on the Living Organism*, which was considered a world reference in this field of study for decades. In the World War II, the need to better protect pilots from the effects of +Gz and −Gz increased. Six human centrifuges were built (one in Canada, one in Australia, and four in the USA). The first UK human centrifuge was built in 1954.

Aviation medicine and space life sciences have also benefited from the use of centrifuges for pilot and astronaut (Figure 2.8) training (Akima et al, 2005; Gaffe et al, 1993; Iwase, 2005; Harding and Mills, 1983; Ohira et al, 2004; Smith, 1992). Terrestrial animals, including humans, require

FIGURE 2.7: Centrifuge used to treat mental diseases in the 1800s. Note that the patient is submitted to −Gz (White, 1964).

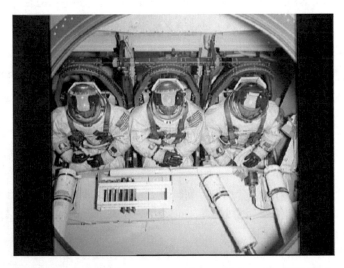

FIGURE 2.8: Astronauts training in the human centrifuge of Johnsville (USA) that replicates the Command Module of the Apollo Program (http://www.nasa.gov).

regular periodic gravitational stimulation to maintain terrestrial physiological functions on Earth or in space. Using a centrifuge, G simulation equivalent to that of Earth can be produced in space. By increasing the levels of G above 1G, such simulation may be more inexpensive, efficient, and effective in preventing the physiological deconditioning in space than many other countermeasures not using artificial gravity. The use of periodic, increased G exposures in space referred to as artificial gravity may offer a possible countermeasure to physiological deconditioning in-flight (Brinkley and Raddin, 1985). Theoretically, artificial gravity can be expected to be more comprehensive in its effect on human physiology, acting on more than one body system as it does, than countermeasure interventions currently used (Burton, 1994).

Several physiological studies have also been conducted in human centrifuges (Burton and Whinnery, 1985; Cardus, 1994). A new one has just arrived at the Institute of Aerospace Medicine of the German Aerospace Centre (DLR), located in Cologne, and was built by the European Space

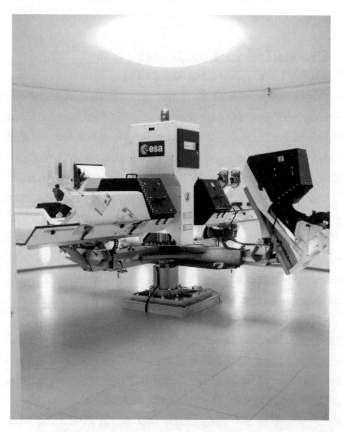

FIGURE 2.9: DLR centrifuge at the Institute of Aerospace Medicine (ESA, 2007)

Agency (ESA). This centrifuge can hold up to four subjects (four arms), and each arm can be tilted in different angles (Figure 2.9).

Centrifugation has also been used widely in the separation of cellular components but only occasionally as a primary environmental stimulus. Edwards and Gray (1977) performed a series of studies looking at the effect of hypergravity simulation on cells and *Escherichia coli*. In the range of +2Gz to +25Gz, auxin transport and geotropic response in coleoptiles (the protective sheath around the embryonic shoot in grass seeds) are increased, and growth seems to be stimulated. Above that, however, damages have been observed both in cells and *E. coli*. From +25Gz to +500Gz, coleoptile growth is reduced and some morphological changes may be seen. At +1000Gz to +2500Gz, root formation in willow cuttings increases. From +1000Gz upward, cytoplasmic stratification occurs and seed germination decreases. Between +200Gz and +15,000Gz, chromosome damage has been observed. Algal cell polarity may be reversed at +5000Gz to +20,000Gz. Above +30,000Gz, the response of some cells to gibberellic acid is halted. Permanent morphological changes in *E. coli* are produced at 110,000Gz. Some plant cells have survived +176,000Gz for 20 h.

2.4   THE MICROGRAVITY CENTRE HYPERGRAVITY EXPERIMENTS

This section presents the conception and development of a centrifuge by Engineer Felipe Prehn Falcão for the performance of hypergravity studies at the Microgravity Centre/FENG-PUCRS, Brazil. This work was published at the *IEEE Engineering in Medicine and Biology Magazine* (Russomano et al, 2007) and was based on a previous study by Russomano et al (2004).

The hypergravity experiments were recorded via a digital camera, and the images acquired were processed for better visualization of the effects of the simulated variation of gravitational force on models of living systems (Coelho et al, 2005) and plant germination (Vieira et al, 2007).

Four different test models were used to simulate the effects of +1.5Gz and +7Gz on human physiology, including the cardiovascular and musculoskeletal systems. One experiment was conducted to evaluate plant germination and growth after intermittent hypergravity exposures to +7Gz, which can demonstrate the usefulness of centrifuges on plant development in extraterrestrial colonies.

The Microgravity Centre centrifuge (Figure 2.10) consists of a motor fixed in a steel base (height of 120 mm, length of 350 mm, and width of 230 mm) supporting a 1200-mm-long arm. The motor is connected to the center of the arm, dividing it into two equal arm lengths of 600 mm, thus allowing two experiments to be conducted at the same time.

The test model is placed at the end of each arm, along with a digital camera for image acquisition during rotation, and an accelerometer, used to indicate the level of acceleration achieved in an experiment.

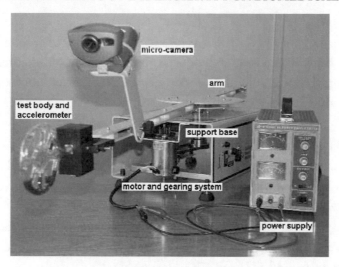

FIGURE 2.10: Microgravity Centre centrifuge with test body at rest (http://www.pucrs.br/feng/microg).

The final gear relation of 5:1 is achieved in two steps. The first gear, attached to the motor, is 23 mm in diameter and is connected to another gear with a diameter of 74.75 mm, providing a ratio of 3.25:1. The second gear is connected by an axis to one with a diameter of 32 mm, which in turn is connected to another gear of 49.6 mm, giving a ratio of 1.55:1. The final ratio of 5:1 is obtained by multiplying 1.55 by 3.25.

Rotation of the system is controlled by a simple DC power supply that is directly connected to the DC motor. The desired rotation per minute (rpm) is established by changing the voltage of the power supply.

The acceleration is measured using an ADXL150 accelerometer. The signal is digitalized through an A/D converter and converted to RS232 communication protocol by an ADuC812. The signal is then sent to a computer via radio Radiometrix TX2 UHF transmitter. A Radiometrix RX2 UHF receiver is attached to the computer to receive the data. The motor is able to accelerate the system up to 36Gz, equivalent to 36 times the force of gravity on Earth.

The G force was calculated using Equations 2.1 and 2.2. The force diagram of the system is represented in Figure 1.11. The weight (W) acting on a test body at rest is balanced by the normal force (N). During rotation, however, a centripetal force (F_{CP}) starts acting on the test body pulling it toward the center of the circle, whereas the inertia keeps the test body in a tangential direction. The resultant force, that is, the apparent weight (W_A) produced is related to the G effect applied on the test body.

$$W_A{}^2 = F_{CP}{}^2 + W^2$$

$$(2.1)$$

$$G = W_A/W \qquad\qquad (2.2)$$

A digital camera with the following specifications was used to acquire the images during rotation: still image capture resolution of 320×240 pixels; video-capturing resolution of 640×480 pixels; and video capture speed of 30 frames per second. The camera was compatible with a PC operating system and Windows 98.

2.5   TEST MODELS

Four different test models (Figure 2.11) representing body systems were used to demonstrate the effects of two levels of hypergravity: a sealed system of tubes filled with a colored liquid to simulate a column of blood (a), three magnetic rings representing the vertebrae (b), a spring (c), and a rubber band (d) with an 8-g mass placed on the free extremity to simulate body's soft tissues (skin, tendons, muscles). After the test model was placed at the end of the centrifuge arm, the system was rotated at 38 and 92 rpm to produce +1.5Gz and +7G, respectively.

2.5.1   System of Tubes Simulating the Column of Blood

At rest, the colored liquid inside the system of tubes was in the lower half of the structure. As the +Gz increased during rotation, the liquid was displaced partially at +1.5Gz and completely at +7.0Gz toward the far end of the system of tubes (Figure 2.12).

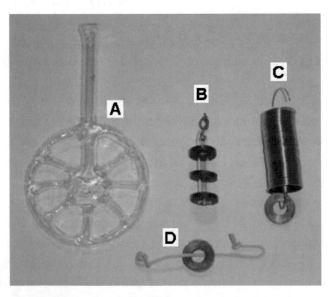

FIGURE 2.11: Microgravity Centre test models (http://www.pucrs.br/feng/microg).

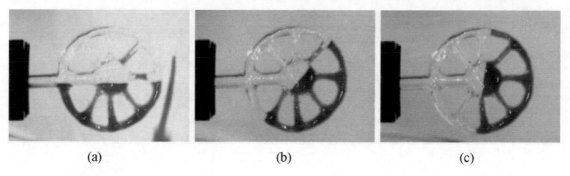

(a)   (b)   (c)

FIGURE 2.12: The system of tubes filled with a colored liquid at 1G (0 rpm) (a), at +1.5Gz (38 rpm) (b), and at +7Gz (92 rpm) (c) (http://www.pucrs.br/feng/microg).

This experiment can be used to represent the column of blood in the human body during different levels of hypergravity exposure. Both the blood in the vascular system and the colored fluid in the capillary system at +1G are attracted to the center of Earth due to the action of gravity. In microgravity, where the resultant force is virtually zero, the blood is uniformly distributed in the human body.

When +Gz force increases, however, due to the centrifuge rotation, the colored liquid is progressively displaced toward the end of the capillary system. This is analogous to what happens to the circulation of a pilot in a fighter aircraft when it suddenly changes direction during a steep maneuver, exposing the pilot to high +Gz acceleration. This force can be sufficient to induce a G-LOC. As explained before, the physiological cause of G-LOC is insufficient perfusion of blood to the brain and subsequent hypoxia. If the onset rate of +Gz acceleration is gradual, the brain hypoxia shows up first as reduced peripheral vision. First, a reduction of the peripheral vision (gray-out) is noted, followed by a loss of the central vision (blackout). Figures 2.12b and 2.12c can be used to schematically represent gray-out/blackout and G-LOC, respectively.

2.5.2   Magnetic Rings Representing Intervertebral Disks

During hypergravity, the magnetic rings were progressively displaced from their resting position at 1G, decreasing the distance between individual rings during rotation (Figures 2.13).

Weightlessness and bed rest, a widely used ground-based microgravity simulation, reduce the mechanical loading on the musculoskeletal system and consequently increase the distance between intervertebral disks (LeBlanc et al, 1994). This effects increases the nutritional diffusion distance and alters the mechanical properties of the spine. The spinal lengthening is thought to cause back pain in astronauts. The opposite, however, occurs during hypergravity exposure, as simulated in Figures 2.13b and 2.13c with the three magnetic rings representing intervertebral disks.

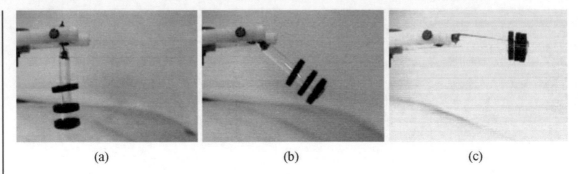

(a) (b) (c)

FIGURE 2.13: Three magnetic rings at 1G (0 rpm) (a), at +1.5Gz (38 rpm) (b), and at +7Gz (92 rpm) (c) (http://www.pucrs.br/feng/microg).

Previous magnetic resonance imaging (MRI) studies have shown that repeated exposure to +Gz forces can cause premature degenerative changes of the cervical spine, as a work-related disease. A study was conducted by Hamalainen et al (1999) on two clinical cases of +Gz-associated degenerative cervical spinal stenosis, a condition in which the distance between intervertebral disks is reduced due to the action of dorsal osteophytes in fighter pilots. Conventional X-rays and MRI were used to demonstrate narrowing of the cervical spinal canal. The first case was complicated by a C6–C7 intervertebral disk prolapse and a congenitally narrow spinal canal. The second case involved progressive degenerative spinal stenosis in the C5–C6 disk space. These two cases suggest that +Gz forces can cause degenerative spinal stenosis of the cervical spine. Flight safety may be jeopardized if symptoms and signs of medullar compression occur during high +Gz stress.

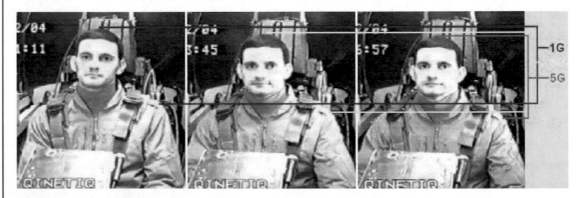

FIGURE 2.14: Test subject during hypergravity exposure at the Royal Air Force human centrifuge in Farnborough, England. Note the progressive decrease in height from +1G to +5Gz (http://www.pucrs.br/feng/microg).

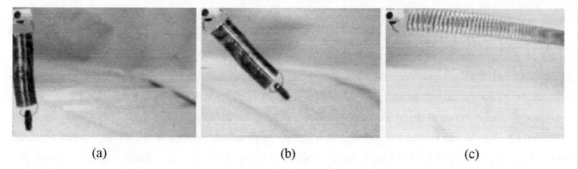

(a) (b) (c)

FIGURE 2.15: Spring with a mass at 1G (0 rpm) (a) at 1.5Gz (38 rpm) (b), and at 7Gz (92 rpm) (c) (http://www.pucrs.br/feng/microg).

Figure 2.14 shows the decrease in height of a test subject inside the gondola of the Royal Air Force human centrifuge in Farnborough, England, during exposures to +3.6Gz and +5Gz.

2.5.3   Spring and Rubber Band With an Attached Mass Modeling Muscle Function

The spring (Figure 2.15) and the rubber band (Figure 2.16) with an attached mass were increasingly stretched at +1.5Gz and +7Gz.

Numerous observations suggest the importance of gravitational force in regulating muscle mass. Centrifugation is believed to be useful for preventing muscle functional and structural loss under microgravity conditions. Ohira et al (2004) studied the effect of hind limb suspension in rats, used to simulate microgravity, or centrifugation at +2Gz between postnatal day 4 and month 3 and

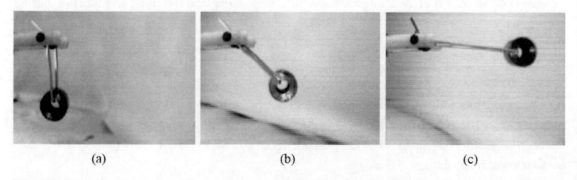

(a) (b) (c)

FIGURE 2.16: Rubber band with a mass at 1G (0 rpm) (a), at 1.5Gz (38 rpm) (b), and at 7Gz (92 rpm) (c) (http://www.pucrs.br/feng/microg).

after a 3-month recovery period back in 1G on the characteristics of muscle. Pronounced growth inhibition was induced by unloading but not by +2Gz force. It is suggested that the development and/or differentiation of soleus muscle fibers are closely associated with gravitational force. The data showed that gravitational unloading during postnatal development inhibits the myonuclear accretion as indicated by the subnormal numbers of both mitotic active and quiescent satellite cells. Although the fiber formation and longitudinal fiber growth were not influenced, cross-sectional growth of muscle fibers was also inhibited together with lesser myonuclear DNA content per unit volume of myonucleus. Unloading-related inhibition was generally normalized after the recovery.

Akima et al (2005) studied the effect of intensive cycle training with artificial gravity on the maintenance of muscle size during bed rest. Ten healthy men were divided into 2 groups: a countermeasure group (BR-CM; $n = 5$) and a control group (BR-Cont; $n = 5$). The BR-CM subjects undertook intensive cycle training (to 90% of maximum heart rate) with short-arm centrifuge-induced artificial gravity on alternate days during 20 days of bed rest. Muscle volume of the thigh and maximum voluntary contraction (MVC) during isometric knee extensions were measured before and after bed rest. Muscle functional magnetic resonance imaging (mfMRI) and electromyogram (EMG) of the quadriceps femoris were obtained during submaximal knee extension exercises at a load of 30% MVC.

The findings showed that the volume of the total thigh muscles was maintained in the BR-CM group (−1%), whereas it was not maintained in the BR-Cont group (−9%, $P < .05$). MVC decreased in the BR-CM (7%) and BR-Cont groups (23%). EMG activity in the BR-CM group after bed rest was significantly lower than before; however, no significant change was found in the BR-Cont group. There were no significant changes in the resting and exercised mfMRI signals in either the BR-CM or BR-Cont groups. These results suggest that intensive cycle training with hypergravity maintained the size of human skeletal muscles during bed rest.

In conclusion, the Microgravity Centre centrifuge, used to perform different hypergravity experiments, succeeded in modeling cardiovascular, musculoskeletal structure, and function during hypergravity.

2.5.4 *Eruca sativa* Mill (Rocket Plant)

A plastic container (60 mm diameter and 62 mm height) with 44 g of black sand (Humosolo, humus type) and 10 seeds of *E. sativa* Mill (KAD type: humidity-proof) was placed at each end of the centrifuge arms and rotated at +7Gz (92 rpm) from 8 A.M. to 5 P.M. daily (a total of 9 h/day) for four consecutive days.

The containers were covered with a plastic lid to avoid evaporation drying of the sand by the generated wind. The container was safely secured to the end of the centrifuge arm through a metal pin. It was placed at 90° in relation to the centrifuge arm, remaining at that position at rest. During centrifuge rotation, it moved upward reaching an angle of 0°. One hole of 3 mm was made on the side of each plastic container to allow ventilation. Two similar plastic containers (control) kept open in the same room served as control. Room temperature was set at 22°C. Water (0.5 mL) was added to the containers before and immediately after the experiment. The experiment was performed twice to test reproducibility.

Rocket seeds are expected to germinate within 4 to 7 days. It was observed that the seeds exposed to +7Gz germinated in 3 days as compared to 4 days for the control seeds. Individual plant height could not be obtained during the experiment days because measurements had to include the roots. The growing plants were removed at the end of the experiment and then measured with a paquimeter from root to top. Two experiments were performed under the same conditions.

Mean plant height at +7Gz in the first experiment was 3.2 versus 1.9 cm of the control. However, a statistical test could not be performed because of the low number of germinated seeds of the control ($n = 3$) in relation to the seeds exposed to hypergravity ($n = 9$). Figure 2.17 shows the difference in plant growth at 1G and +7Gz intermittent exposure. In experiment 2 ($n = 14$), mean plant growth of 2.2 cm at +7Gz was significantly higher than the plant growth at 1G (control) of 1.8 cm ($P = .02$).

FIGURE 2.17: Difference in plant growth at 1G and +7Gz intermittent exposure (http://www.pucrs .br/feng/microg).

The result of above-average growth of the plants after being exposed to hypergravity simulation, as well as the increased number of germinated seeds found at +7Gz in experiment 1, motivated a series of studies related to the effect of hypergravity on plant growth (Vieira et al, 2007).

The Microgravity Centre centrifuge was then adapted to be able to hold 12 containers of soil and *E. sativa* Mill seeds at the same time, all rotating at +7Gz from 8 A.M. to 5 P.M. for four consecutive days. A round plastic structure (600 mm in diameter) was built to accommodate the plant containers. It was adapted to top of the centrifuge steel base in a substitution for the previous two-arm structure. Figures 2.18 and 2.19 show an schematic view of the round plastic structure and the container used during the experiment, respectively.

Figure 2.20 shows the experiment being performed. Note the centrifuge with the round structure during rotation ($n = 12$) and the eight plastic containers placed (control at 1G) in front of the rotating centrifuge.

A series of experiments were conducted, and the results confirmed the previous findings. During 9 h of intermittent exposure to +7Gz, for example, the plants grew 8.44 ± 1.43 cm ($n = 12$). The control at 1G had a growth of 3.76 ± 1.21 cm ($n = 8$) ($P = .00$) (Vieira et al, 2007).

These results motivated the patent of the process of plant growth in hypergravity by the Microgravity Centre/FENG-PUCRS, Brazil.

A theory that might explain the significant growth of the *E. sativa* Mill after 4 days of intermittent hypergravity exposure is based on the influence of the auxins on plant growth.

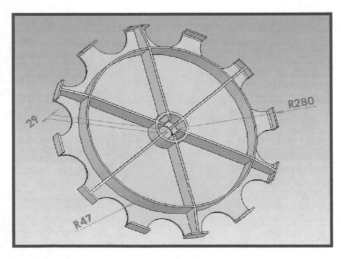

FIGURE 2.18: Schematic view of the round structure used to support the plant containers ($n = 12$) (http://www.pucrs.br/feng/microg).

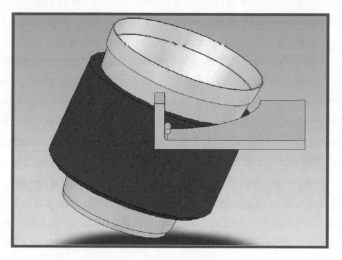

FIGURE 2.19: Schematic view of the plant container (http://www.pucrs.br/feng/microg).

FIGURE 2.20: The *E. sativa* Mill experiment (http://www.pucrs.br/feng/microg).

The plant hormone auxin, a term derived from a Greek word that means "to grow," is characterized by its capacity to induce the elongation of cells in the sub-apical region of branches. Auxin also affects physiological processes, including phototropism, geotropism, fruit development, and gender expression. However, cellular elongation is one of the most important effects of the Auxin action. An important step of elongation is the acidification of the cellular border. This is generated by an electrochemical gradient that leads to proton secretion through the plasmatic membrane, promoting the acidification of the cellular wall. This leads to an increase in enzymatic activity, which promotes the malleability of the cell edges that allows cellular elongation. Then, the osmotic pressure forces the entrance of water into the cell, causing the expansion of the cell.

Three new ESA facilities will be available for biological experiments in Space, Biopack on the Space Shuttle and two instruments on the International Space Station: BIOLAB in the European "Columbus" Laboratory and the European Modular Cultivation System (EMCS) in the US Lab "Destiny."

The experiments are housed in standard experiment containers, allowing either research in microgravity or acceleration studies with variable G levels, when mounted on the centrifuges. Whereas Biopack provides only thermal control, BIOLAB and EMCS supply each container with a dedicated atmosphere (controlled CO_2 and O_2 concentrations, and relative humidity and trace gas removal): EMCS also contains fresh-water and waste-water reservoirs on its rotors. Power and data lines are available in all the described facilities. Highly automated systems, like BIOLAB's Handling Mechanism and Analysis Instruments, support the tele-science concept and help reduce crew time in orbit. A BioGlovebox with its support instruments offers unique research facilities in Space. The feasibility of offering experiment hardware inside the containers has been studied by ESA for several kinds of experiment support equipment that could potentially be used for research in developmental biology; for example, in experiments with small eggs, with cells and tissues, with small aquatic animals, with insects, and with plants (Brinckmann, 2003).

Centrifuges are particularly important for research in gravitational biology because the inertial forces developed by motion can be combined with gravitation to produce gravitational fields other than Earth's gravity. In orbiting satellites, centrifuges can provide an on-board 1G environment. They can also be very useful for Mars and Moon colonies to counteract the effects of hypogravity or microgravity on human and animal physiology and plant growth (Young, 1999).

· · · ·

CHAPTER 3

The Effects of Microgravity on Biomedical Experiments

3.1 GROUND-BASED MICROGRAVITY SIMULATION

The importance of an adequate understanding of the physiological responses to microgravity to assure the health and well-being of astronauts in space has grown since the beginning of manned spaceflight and has motivated a series of biomedical experiments in the Skylab, Space Shuttle, and Mir Space Station programs, as detailed in previous chapters. However, many factors associated with spaceflight activities complicate attempts to delineate the time course of physiological responses to microgravity, including the following:

Sample size—crew sizes have ranged from five to eight astronauts and only two or three of them have been allowed to participate in biomedical experiments. Any attempt to extrapolate from this small number to a larger population is unsatisfactory.

Limited capabilities for scientific observations—biomedical experiments are restricted by operational limitations and the time available during a space mission.

Extensive use of countermeasures—the prophylactic and therapeutic use of a variety of countermeasures has masked the direct effects attributable to microgravity alone on human adaptation to space environment.

Different mission types—frequent changes of mission profiles make direct comparisons between flights difficult.

These limitations to the conduct of biomedical experiments during space missions has led to the widespread development of ground-based microgravity simulations, which have been used to study the human physiological responses to microgravity and to develop effective countermeasures for use in spaceflights.

Among the ground-based simulation techniques are bed rest and bed rest associated with head-down tilt (HDT), water immersion (neutral buoyancy), immobilization, clinostat (2D and 3D) for plants and small animals, and total- and partial-body suspension.

3.1.1 Bed Rest and Bed Rest-Associated HDT

Bed rest associated with HDT is the most widely ground-based simulation technique used to study mainly the cardiovascular responses to microgravity. Studies, however, have contemplated other body systems and functions, such as neuroendocrine system control, psychological behavior, and sleep–wake cycle changes. Bed rest with or without HDT may vary from minutes to months depending upon the study objectives.

It is well known that the cardiovascular disturbances, which begin immediately after the insertion into microgravity, include a major redistribution of blood and extravascular fluids from the lower to the upper part of the body. Studies using multiple girth measurements revealed a total reduction of 1.5–2.0 L in the volume of the lower limbs of astronauts in Skylab IV and subsequent Space Shuttle flights. This redistribution of blood and extracellular fluid, due to the removal of the hydrostatic pressure gradients (headward shift), causes nasal and sinus congestion, puffiness of the eyelids, rounding of the face, engorgement of the superficial veins of the head and neck, and a sensation of fullness in the head. The cardiovascular effects of the shift of blood are increases in cardiac preload, heart size, resting heart rate, blood pressure, and stroke volume.

Upon return to Earth, the hydrostatic gradients produced by standing erect challenge the cardiovascular system as demonstrated by the increase in heart rate, the decrease in mean arterial blood and pulse pressures, the fall in aerobic exercise capacity, and the reduction in orthostatic tolerance in response to tilting upright and/or lower body negative pressure. These cardiovascular responses to the Earth's gravitational force (1G) after an exposure to microgravity are designated *cardiovascular deconditioning*.

The first head-down experiment was carried out at an angle of 4° to assess the effectiveness of countermeasures to be used in the flight of the Salyut orbital station. Kakurin et al (1976) exposed eight subjects to 5 days of bed rest at 0° (supine position), 4°, 6°, 8°, and 12° HDT. The cardiovascular responses to 75° head-up tilt (HUT) and exercise after 5 days of bed rest at these angles of HDT and after 5 days of spaceflight were compared. Clinical symptoms similar to those that occur in microgravity secondary to the headward shift of blood (sensation of blood rushing to and heaviness in the head, nasal congestion, facial puffiness and distension of scleral and conjunctival vessels, and a sensation of fullness in the eyes) were absent in the supine position. These symptoms were reported to be mild, moderate, and intense at 4°, 8°, and 12° HDT, respectively. In fact, the nasal congestion was so severe during 8° and 12° HDT that rhinitis developed in three subjects. The increase in heart rate (mean of 19 bpm) during the 75° HUT test recorded in the seven cosmonauts after 5 days of spaceflight was found to be between the increase in heart rate during the 75° HUT test shown by the subjects who had been exposed to 4° and 8° HDT. A comparative assessment of the ergometry test carried out after the bed rest studies and the 5-day spaceflight showed that post-flight changes in heart rate and oxygen consumption approximated the changes of the subjects who were kept at 4°, 8°, and 12° HDT. This study suggested that bed rest associated with HDT between 4° and 12° elicits the effects of microgravity on the cardiovascular system with greater fidelity than does horizontal bed rest (supine position). However, it was also found that 8° and 12° HDT caused undesirable symptoms. Bed rest with 6° HDT was, therefore, chosen as the best angle to simulate the cardiovascular responses to microgravity, and it has been widely used for this purpose since 1976.

3.1.2   Tilt Tables

The Microgravity Centre/FENG-PUCRS, Brazil, and King's College London, UK, have been conducting joint experiments using HDT as a method of microgravity simulation.

The KCL tilt table is a hydraulic–pneumatic tilt table (2 m long and 55 cm wide) used to place the subject at various angles to the horizontal, including 6° head-down, 70° head-up, and 0° (supine) (Figure 3.1). The tilt table can be rotated from −6° to +70° to the horizontal by a hydraulic jack, driven by compressed air. The angle at which the subject is positioned is determined by a plum line attached to a protractor, which is secured to the table. The gas and oil control valves are set to provide transition from +70° to horizontal and from −6° to +70° in 5 and 8 s, respectively.

The tilt table is fitted with an adjustable webbing harness comprising five straps and a quick-release box, which secures the subject during rotation. The lap straps of the harness should be

FIGURE 3.1: Subject during an HDT experiment—KCL tilt table (http://www.pucrs.br/feng/microg).

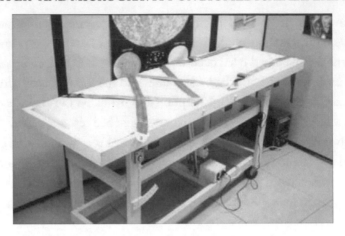

FIGURE 3.2: Microgravity laboratory tilt table (http://www.pucrs.br/feng/microg).

positioned over the subject's iliac crests. The weight of the relaxed subject is supported by either (1) a foot rest and an 8 × 8 × 20-cm metal block is used under the center of subject's feet whenever the 70° head-up position is assumed, to avoid muscle contraction and its effects on the venous return from the lower limbs, or (2) a padded saddle.

The Microgravity Centre tilt table is an electrically controlled tilt table (1.95 m long, 0.64 m high, and 0.6 m wide) (Figure 3.2).

The inclination of the tilt table ranges from 0° (supine position) to −20° (Figure 3.3). The angle at which the subject is positioned is determined by scale placed in the front part of the tilt

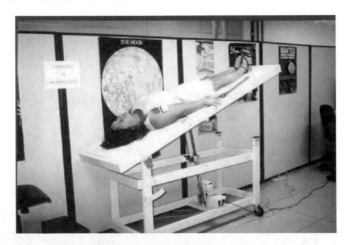

FIGURE 3.3: Subject during an HDT experiment—microgravity laboratory tilt table (http://www .pucrs.br/feng/microg).

table that ranges from 0° to 25°. The electrical control allows the movement to be performed at 1° interval.

A harness used by commercial pilots is attached to the tilt table to promote safety and comfort for the subject during microgravity simulation experiments.

One of the joint studies between KCL and the Microgravity Centre aimed to determine the angle of HDT that best simulates the increase in the intraocular pressure (IOP) after 15 min in microgravity. The redistribution of fluids from the lower to the upper part of the body during a space mission increases IOP with a peak value in the early phase after reaching microgravity. Studies conducted during the first 15 min of German Spacelab D1 and D2 missions showed a 92% (Schwartz et al, 1992) to 114% (Draeger et al, 2003) rise in IOP pressure compared to the baseline values.

In this study, the IOP of eight subjects was measured by a portable tonometer on five separated occasions: sitting upright (baseline) and at the end of 15 min of four different randomized angles of HDT (6°, 12°, 18°, and 34°). The IOPs (mean ± SE) for right and left eyes combined ($n = 16$) were 12.6 ± 0.5 mmHg at sitting position, 17.0 ± 0.7 mmHg at −6°, 18.9 ± 0.7 mmHg at −12°, 19.6 ± 0.7 mmHg at −18°, and 27.0 ± 1.5 mmHg at −34°. The results showed that the IOP significantly increased during HDT regardless the angle of tilt used ($P < .05$). The angle of 34° of HDT, however,

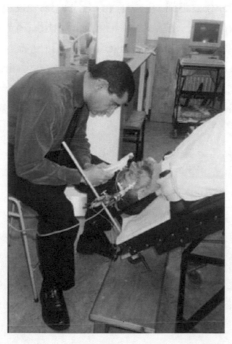

FIGURE 3.4: Simultaneous measurements of IOP and venous pressure of the forehead during 34° of HDT—KCL tilt table (http://www.pucrs.br/feng/microg).

enhanced the IOP by 110% and was accepted as the angle that best simulates the acute increase in IOP that occurs in microgravity. These findings demonstrate that the 6° HDT position, a commonly used method of simulating physiological changes secondary to microgravity exposure, did not produce the rise in IOP that has been found during space missions (Russomano et al, 2001b).

This study led to a second joint research between KCL and the Microgravity Centre with the aim to determine the relationship between IOP and forehead venous pressure produced by 15-min exposures to HDT. The IOP and forehead venous pressures of four subjects were measured by a portable applanation tonometer and a pressure transducer connected to a needle inserted in a forehead vein, respectively. Pressures were recorded at intervals during 15-min exposures at 0° (supine), 17°, and 34° HDT (Figure 3.4) (Russomano et al, 2001a).

The IOPs (mean ± SE) for right and left eyes combined ($n = 8$) and the forehead venous pressure ($n = 4$) were 14.7 ± 0.5 and 5.2 mmHg at 0°, 20.7 ± 0.6 and 13.0 mmHg at 17° HDT, 25.5 ± 0.3 and 19.3 mmHg at 34°, and 13.5 ± 0.8 and 4.0 mmHg at 0° (recovery), respectively, as shown in Figure 3.5.

These results indicate that IOP and forehead venous pressures simultaneously increased during HDT ($r = 0.99$) regardless of the angle of tilt used ($P < .05$), and both returned to their control values in the supine position. Future joint studies are planned to evaluate peripheral venous pressure of the forehead in microgravity (parabolic flights) to clarify the mechanisms involved in the rise of IOP due to weightlessness exposure, which may differ from those found during HDT.

Special facilities have been built at space agencies for long-term HDT experiments because these studies require subjects to be in bed rest for days, weeks, or even months. The Schalfen laboratory at the German Aerospace Center in Cologne, Germany, is a good example of a long-term ground-based microgravity simulation facility and is shown in Figures 3.6 and 3.7.

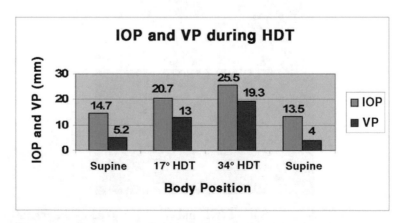

FIGURE 3.5: IOP and venous pressure during different HDT angles (Russomano et al, 2001a)

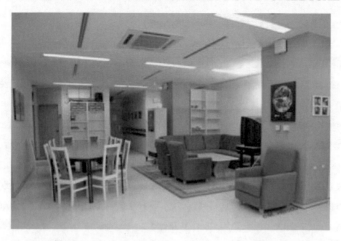

FIGURE 3.6: Schalfen laboratory at the German Aerospace Center in Cologne, Germany. Recreation and meeting area (http://www.dlr.de).

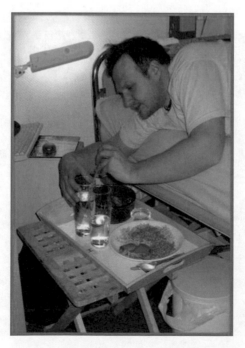

FIGURE 3.7: Subject eating during an HDT experiment (http://www.dlr.de)

3.1.3   Water Immersion (Neutral Buoyancy)

Neutral buoyancy is a condition in which a physical body's mass equals the mass it displaces in a surrounding medium. This negates the effect of gravity that would otherwise cause the object to sink. An object that has neutral buoyancy will neither sink nor float. Instead, it will remain at its current level in the medium that surrounds it.

Since the mid-1960s, the Johnson Space Center/NASA Neutral Buoyancy Facility has been an invaluable tool for testing procedures, developing hardware, and training astronauts. It simulates reduced gravity for the astronaut to practice future in-flight procedures, such as extravehicular activities, and to work through simulation exercises to solve problems that the astronaut can encounter in-flight. The neutral bouyancy laboratory (NBL) was sized to perform two activities simultaneously, each of them using mockups sufficiently large to produce meaningful training content and duration. It is 202 ft. in length, 102 ft. in width, and 40 ft. in depth (20 ft. above ground level and 20 ft. below) and holds 6.2 million gallons of water (Figures 3.8 and 3.9).

The Russian Neutral Buoyancy Facility is located at the Gagarin Cosmonauts Training Center (GCTC). It consists of a larger hydrolaboratory capable of accommodating a 20-ton space station module. The pool has a depth of 12 m, diameter of 23 m, and volume of 5000 m^3 (Figure 3.10).

FIGURE 3.8: Johnson Space Centre/NASA Neutral Buoyancy Facility (http://www1.jsc.nasa.gov).

FIGURE 3.9: Astronaut training at JSC/NASA Neutral Buoyancy Facility (http://www1.jsc.nasa.gov).

FIGURE 3.10: Russian Neutral Buoyancy Facility at GCTC (http://www.nasa.gov).

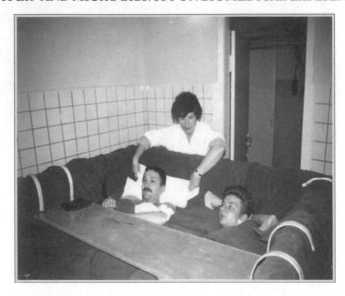

FIGURE 3.11: Head-out head immersion (Russian Federal Space Agency; http://www.roscosmos.ru).

During training in a NBL, suited astronauts must experience no buoyant force and no rotational moment about their center of mass. The suits worn are down-rated from fully flight-rated extravehicular mobility unit suits, like those in use on the Space Shuttle and International Space Station. The presence of drag is one of the negative aspects of water immersion as a microgravity simulator. Generally, this is overcome by doing tasks slowly inside the water, to minimize the effect of drag on the exertion required to complete a task. Another downside is that astronauts are not weightless within the suit, and thus, suit sizing is critical.

Neutral buoyancy and head-out water immersion (Figure 3.11) are not considered analogues of a microgravity simulation to study human physiology due to the effects of the external hydrostatic pressure on the resilient body tissues and gas-containing body cavities. This results in physiological alterations in the cardiovascular and respiratory systems, and in gastrointestinal tract. The increase in venous return is estimate to cause a 32% to 66% increase in cardiac output and the right to left shunting because of a combination of pulmonary pooling, cephalic movement of the diaphragm, and compression of the chest wall. The central shift in blood volume is perceived by the body as hypervolemia, resulting in increased diuresis, natriuresis, and kaliuresis.

3.1.4   Immobilization

Total- or partial-body immobilization has been used as a simulator of microgravity for the musculoskeletal system in humans and animal models. The lack of weight bearing, as occurs in space, is associated with reductions in strength and mass of skeletal muscle. In the last few years, a human model to simulate unloading of lower limb skeletal muscles that occurs in microgravity has been used. This model was essentially adopted from the rat hind-limb suspension technique and can also be used to study the hind-limb suspension-induced bone loss.

3.1.5   Body Suspension Devices

Body Suspension Device System (BSDS) has been used to study human physiology during either microgravity or hypogravity simulations. The BSDS developed by the Microgravity Centre was used to study, for example, the walking pattern in ground-based martian and lunar simulation experiments, in which the length, duration, speed of the stride, cadence, and duration of the stance phase have been evaluated (Figure 3.12) (Newman, 2000).

The Cleveland Clinic Foundation and Penn State University have built the Zero-Gravity Locomotion Simulator (ZLS). The ZLS is mounted vertically in a free-standing frame and includes padded straps that support a runner under the head, torso, arms, and legs (Figure 3.13). In

FIGURE 3.12: Subject partially suspended during Mars simulation (http://www.pucrs.br/feng/microg).

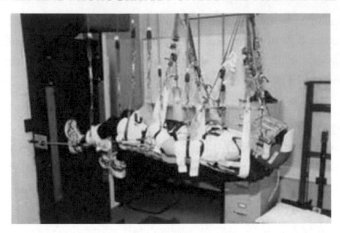

FIGURE 3.13: Subject exercising on the ZLS treadmill (http://www.csuohio.edu).

this position, there's no gravitational force between the runner and the machine. Like the treadmill used on the space station, the ZLS has a harness with motors and cables to pull the runner toward the treadmill belt, exerting force on the bottom of the runner's feet.

3.1.6 Barany's Chair

Rotatory chairs have been widely used to study the vestibular system and to simulate motion sickness (Figures 3.14). They are also called *Barany's chair*, after the Austrian otologist who won the Nobel Prize for Physiology or Medicine in 1914 for his work on the physiology and pathology of the vestibular (balancing) apparatus of the inner ear. Robert Barany graduated in medicine from the University of Vienna in 1900. After studying at German clinics, he became an assistant at the ear clinic of the University of Vienna and, in 1909, a lecturer on otological medicine. He devised new tests for detecting vestibular disease and for examining activities of the cerebellum and their relation to disturbances of equilibrium. Barany served in the Austrian army in World War I and was taken prisoner by the Russians in 1915. He was a prisoner of war when he was awarded the Nobel Prize that year. From 1917 until his death, he taught at Uppsala University, where he was head of the ear, nose, and throat clinic.

Motion sickness is a fairly common problem in spaceflights. It affects approximately 70% of the astronauts during the first 72 h of a space mission. Motion in flight, at sea, in a car, or in a Space Shuttle generates patterns of sensory input that conflict with those patterns based on terrestrial experience. The brain is unable to interpret this conflict and motion sickness results. The otolith organs, which lie within the utricle and saccule, are responsible for the perception of linear accel-

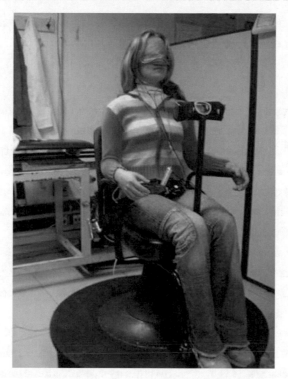

FIGURE 3.14: Subject in the Microgravity Centre rotatory chair at rest (http://www.pucrs.br/feng/microg).

eration when it is greater than 0.1 m/s². The semicircular canals provide reliable information about angular accelerations of the head and may be regarded as natural accelerometers working as three matched pairs. Sustained change in the angular velocity of the head greater than approximately 3°/s is detected by the canals in the plane of movement, and its magnitude and direction are signaled to the brain. Movement will be detected by the canals only for as long as there is a suprathreshold acceleration or deceleration. Once constant velocity is reached, the signal will decay, although movement is continuing.

The Microgravity Centre has developed an electrically controlled rotatory chair to study the vestibular system, motion sickness, and the effect of different drugs, especially scopolamine, on the prevention of its signs and symptoms. The Microgravity Centre rotatory chair was built according to the following specifications: rotation range from a minimal of 10 rpm to a maximum of 30 rpm; maximal accidental load of 100 kg; no nominal variation of the acceleration during rotation. A DC motor (24 V) was placed on the axis of the chair to promote its rotation. The nominal rotation of the motor is at 3000 rpm. A reducer was used to decrease it by a factor of 15:1. The transmission of the movement between the motor and the chair is given by two trapezoidal canal pulleys and an A29 model chain with a reduction of 6:1.

3.1.7   3D Clinostat

The practice of rotating plants around a horizontal axis as a means of studying plant geotropism dates back to the late 1700s. Devices that accomplished this mechanical manipulation have been called *clinostats* since 1882. As discussed before, on Earth, real microgravity conditions can be produced by a free fall from a drop tower or by parabolic flights of airplanes, for example. However, the duration of microgravity obtained by these methods is too short for plants or cells to exhibit obvious changes in growth and development. The simulated microgravity environment created on Earth within a clinostat can be used without time restriction.

Clinostats can be slow rotating or fast rotating with one or two axes. The slow-rotating clinostats, which are the scope of this section, must rotate with a constant angular velocity, which has to be sufficiently small (2–4 rpm) and the plant or cell sufficiently well centered over the axes of the clinostat to avoid centrifugal effects.

Clinostats, however, cannot fully reproduce the concurrent lack of structural deformation, displacement of intercellular components, and reduced mass transfer in the extracellular fluid that occurs in actual weightlessness. It is evident that gravity-specific alterations in plants, cells, and organisms can be proven beyond doubt only by experiments performed during space missions, which have limited access and high cost.

For suspension cell cultures, a state termed *functional microgravity* can be achieved using a clinostat by rotating a container completely filled with liquid at a constant velocity. After a brief startup period, the rotational velocity of the container wall is transferred radially inward until no

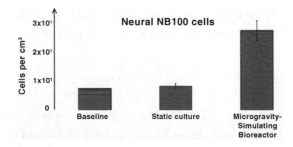

FIGURE 3.15: Neural NB100 cells increase by 3.3-fold for neural NB100 cells [microgravity-simulating bioreactor (MSB): $2.76 \pm 0.35 \times 10^4$ cells/cm^3; static cultures (SCS): $0.83 \pm 0.09 \times 10^4$ cells/cm^3 ($P < .001$; n = 6)].

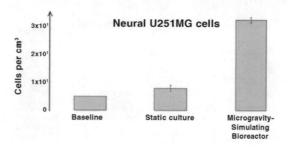

FIGURE 3.16: Neural U251MG cells increase by 4.1-fold for neural U251MG cells [MSB: $3.21 \pm 0.09 \times$ cells/cm^3; SCS: $0.78 \pm 0.27 \times 10^4$ cells/cm^3 ($P < .001$; n = 6)].

relative fluid motion exists and the fluid rotates as a rigid body with the suspended particles being randomly distributed while following small circular paths.

A 3D clinostat was developed by the engineer Felipe Prehn Falcão from the Microgravity Centre and was validated by the Stem Cell Group of Kingston University London, which used four different types of human cancer cells and cord blood stem cells (CBSCs). After rotation for 19 h at 37°C, 5% CO_2 humidified atmosphere, the 3D clinostat significantly improved proliferation potential of all tested cell populations when compared to static cultures, as shown in Figures 3.15, 3.16, 3.17, and 3.18. After only 5 days, high-definition microscopic analysis revealed that all CBSCs adhered and expanded onto the BD™ 3D Collagen composite scaffolds and cross-developed into hepatocyte-like cells on stimulation (Forraz et al, 2004).

A new version of the Microgravity Centre clinostat was used in a joint research with the Department of Pharmacy, King's College London, in 2006 (Figure 3.19) (Santos et al, 2006). This new 3D clinostat has two turning axis, a maximum external dimension of 47 by 47 cm and the capability

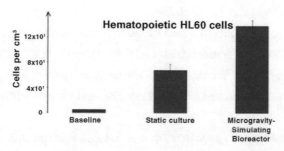

FIGURE 3.17: Hematopoietic HL60 cells increase by 3.3-fold for neural NB100 cells [MSB: $2.76 \pm 0.35 \times 10^4$ cells/cm^3; SCS: $0.83 \pm 0.09 \times$ cells/cm^3 ($P < .001$; n = 6)].

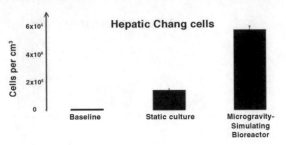

FIGURE 3.18: Hepatic Chang cells increase by 4.1-fold for hepatic Chang cells [MSB: 57.1 ± 0.25 × cells/cm³; SCS: 13.92 ± 0.25 × 10⁴ cells/cm³ ($P < .001$; n = 6)].

for testing eight microgravity diffusion chambers with samples inside at the same time. The rotation speed of both axes is fixed at 1.6 rpm. Human bronchial epithelial Calu-3 cells were used to model the airway epithelium. Cells were seeded on 24-well polyester Transwell® cell culture inserts at a density of 10^5 cells/cm². Cells were transferred after 24 h to an air–liquid interface in specifically designed clinostat diffusion chambers and cultured either under gravity ($n = 4$) or under microgravity simulation in the 3D clinostat ($n = 4$) for 20 days to investigate the effect of microgravity on cell growth. Transepithelial electrical resistance (TER) was used to monitor the development of the cell layer permeability barrier. To investigate epithelial cell layer function in microgravity, cells were cultured under standard culture conditions for 14 days, then exposed to microgravity simulation for 24 or 48 h ($n = 4$) and evaluated by TER and mannitol flux (as a measure of solute permeability) measurement. Results showed that cells grown under gravity or in microgravity simulation formed layers exhibiting TER values higher than 350 $\Omega \cdot$ cm² after 12 and 20 days in culture, respectively. Confluent Calu-3 monolayers transferred to the diffusion chambers maintained a high TER after 24 or 48 h in the 3D clinostat. These findings indicate that a "clinostat diffusion chamber" is suitable for growing monolayers of epithelial cells and for investigating their permeability in vitro during microgravity simulation in the 3D clinostats.

Space tourism is a recent phenomenon of space travel for the purpose of personal pleasure, presently only affordable to wealthy individuals and corporations. Among the primary attractions of space tourism is the uniqueness of the experience, the thrill and awe of looking at Earth from space, the idea of the experience as an exclusivist status symbol, and various advantages of weightlessness (Nagatomo, 1994).

Because of the sprouting of space tourism and the fact that spaceflights have become a more prevalent endeavor, an increase in the number of studies on the effects of microgravity on organ function, cellular development, and bacterial growth is needed (Lu, 2002). Microgravity per se is

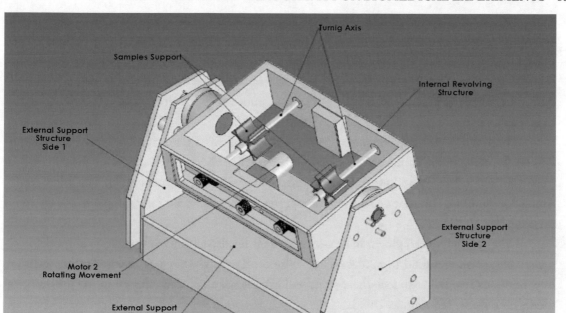

FIGURE 3.19: Schematic view of the Microgravity Centre 3D clinostat (http://www.pucrs.br/feng/microg).

associated with a number of fundamental physiological changes both on organ function and on cellular levels. Documented alterations in human physiology during space missions encompass lower than normal systemic hormone levels and decreased erythrocyte mass (Strollo, 1999; Alfrey et al, 1996). Previous studies have also clearly shown that low-gravity and microgravity environments provoke an intense bone and muscle atrophy (Bungo et al, 1985; Bungo et al, 1987). Weightlessness suppresses many lymphocyte functions on which immunity relies, such as cell locomotion and antigen expression (Sastry et al, 2001). Indeed, real and simulated microgravity decrease in vitro responsiveness to mitogen-mediated stimulation (Taylor, 1993).

The present study was designed to evaluate changes that the immunological system may suffer during an exposure to simulated microgravity, especially on cellular proliferation and viability. Because of the difficulty of having access to a real microgravity environment, it was necessary to develop a tool capable of simulating microgravity on Earth for cells, named *clinostat*, which can be used for unlimited periods.

The 3D clinostat is a microgravity simulator that is based on the principle of gravity vector averaging. Gravity is a vector because it has magnitude and direction. During an experimental run in a two-axes 3D clinostat, the sample's position with regard to the Earth's gravity vector direction is constantly changing. The sample may experience this as a zero-gravity environment. In the clinostat, microgravity simulation is obtained by continuous random changes of orientation, relative to gravity's vector.

The gravity's vector is generated by a combination of two different movements that can simulate results comparable to the effects of true microgravity. The changes are faster than the object response time to gravity and never have a constant direction for a specific time. The microgravity simulation within a clinostat depends on the speed and the distance of the sample to the center of rotation.

Clinorotation, a cell grown in a vessel with zero headspace and slow rotation around a horizontal axis, is one method to model microgravity on Earth. Under such conditions, the cells are in continuous free fall with low-shear stress and experience a constant change in the gravitational vector (Unsworth and Lelkes, 1998).

The clinostat, however, is one of many devices used to simulate the microgravity environment on Earth. Other equipments are random positioning machine, rotating-wall vessel, and free-fall machine. Table 3.1 presents a summary of these microgravity simulator devices with their specific principle for achieving weightlessness, major assets, and pitfalls (Manti, 2006).

In the late 19th century, the clinostat was originally used to find out why plant roots appear to grow predominantly toward the center of the Earth. Over the last 2–3 decades, slow- and fast-rotating 2D and 3D clinostats have been used to assess cellular adaptation to this environment. A cell culture is placed in a spin module of the clinostat platform and its rotation is set empirically (2–3 rpm). The clinostat is then allowed to run for a specified period (hours to days) after which the cultures are removed and assayed for specific properties, such as cell growth, size and shape, distribution of receptors, integrity of the cytoskeleton, or gene expression (Audus, 1962; Klaus, 2001).

A 3D clinostat was developed by the Microgravity Centre/PUCRS, Brazil, to simulate microgravity on Earth. The Microgravity Centre clinostat was the first one conceived, projected, and developed entirely in Brazil. A series of validation studies were conducted (Forraz et al, 2004; Santos et al, 2006). The main objective of the study presented here was to evaluate changes in the immune system through the evaluation of the effect of simulated microgravity on cellular proliferation and viability. It also aimed to validate the new version of the Microgravity Centre 3D clinostat as a tool capable of simulating microgravity (Martinelli et al, 2007).

The 3D clinostat developed by the Microgravity Centre/PUCRS, Brazil, has two turning axes (100 mm long each) where four samples can be attached to each one and tested at the same time. The rotation speed of both axes was fixed at 1.6 rpm, which is equivalent to 0.00168 Gy/s (angular velocity).

TABLE 3.1: Synopsis of commonly used devices for achieving microgravity on Earth (Manti, 2006)

DEVICE	OPERATING PRINCIPLE	MAJOR ASSETS	MAJOR PITFALLS
Clinostat	Axial rotation	Quiescent fluid environment	No gas exchange
Random positioning machine	Directional randomization	Quiescent fluid environment	No gas exchange
Rotating-wall vessel	Solid body-rotation	Low shear formation	No g-jitter
Free-fall machine	Free fall over limited time	Free-fall conditions	Mechanical stress
Centrifuge free-fall machine	Free fall coupled with pulsed centrifugation	Partial gravity allowed	Reduced time
Chemiostat	Dynamic neutralization of sedimentation velocity	Small dependence on cell aggregate size	Bubble formation
Neutral buoyancy	Static neutralization of sedimentation velocity	Small dependence on cell aggregate size	Low cell growth and metabolism

The solid structure of the 3D clinostat was premodeled in the SolidWorks software. This is a 3D computer-aided design program that runs on Microsoft Windows platforms, developed by SolidWorks Corporation, a subsidiary of Dassault Systèmes. S. A. SolidWorks uses a parametric, feature-based approach that creates models and assemblies.

Based on the 3D clinostat SolidWorks design, a solid structure with three main parts was developed, each one made of 8-mm-thick expanded polyvinyl chloride (PVC), that provides good stability and corrosion resistance (Figure 3.20).

The internal part of the Microgravity Centre 3D clinostat is a 288 × 209-mm rectangular structure that supports the samples' containers. The two sample containers are made of nylon, which have four clip-like holders made of 32-mm PVC tube. Each one features a four-clip array surrounding the rotating axis. Therefore, eight samples of cell culture can be exposed to microgravity simulation at the same time. The external support structure is an upside-down V-shaped structure with a size of 383.5 × 220 mm. The distance between the center of the microgravity diffusion chamber and the center of the rotational axis is 25.5 mm.

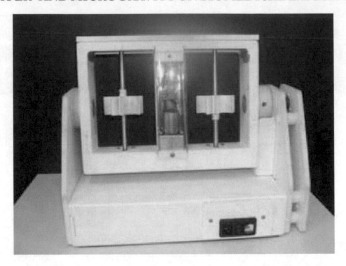

FIGURE 3.20: Microgravity Centre 3D clinostat (http://www.pucrs.br/feng/microg).

For the rotational movements, two 3.8-rpm, 24-V DC motors were used. These motors were coupled to gearing systems. To provide a revolving movement of the whole internal structure, a pulley timing belt system was used to transfer the rotational movement from the motor that is placed underneath the external support structure. For rotational movement of the sample, another pulley timing belt system was used with both rotating axes attached with a pulley and then, via two timing belts, were turned by the same pulley on the motor axes.

A self-switching power supply of 12 V, 1.5 A was used. The revolving motor was connected directly to the power supply. For the rotational motor, two sets of brush systems were used. The reason for the parallel system is to avoid the risk of poor contact or variation in the power supply for this motor.

A graphical user interface for remote control was developed, based on Borland Delphi Compiler, to improve the 3D clinostat with an electronic interface system. The Borland Delphi is a rapid application development tool that supports the Delphi programming language (Object Pascal) and C++ for the 32-bit Microsoft Windows platform, and Delphi and C# for the Microsoft .NET platform.

Humidity and Temperature Sensors. The 3D clinostat operates inside a climatized incubator where the temperature was set to 37°C and humidity of 95%. The Sensirion SHT71 sensor for temperature and humidity was chosen to keep record of the incubator environment variables.

The SHT71 has digital communication, which avoids interference from the environment variables during signal readings. The readings were made in 14-bit resolution for temperature and 12-bit resolution for humidity, considered a sensor with an appropriated resolution for biomedical experiments. The communication is made through a protocol with seven phases: start transmission, sending of address and data, arrival acknowledge signal, standby conversion, receiving of the four most significant bits, received acknowledge signal, and the receiving of the eight less significant bits.

Microcontroller. The controller used for the digital functions was the microcontroller model MSP430f149 from Texas Instruments. The functions that this controller plays are characters display configuration, RS232 serial communication, pulse width modulation (PWM), signal modulation, and the reading and configuration of the temperature and humidity sensor.

During the tests, a development board was used, designed and manufactured by the Microgravity Centre, which allowed the access to all functions of the microcontroller. The board also has a RS232 serial communication converter, characters display output, and the microcontroller serial and parallel recording circuits.

Radiofrequency Transceiver. The remote communication system developed by the Microgravity Centre uses the Wenshing TRW24G model radiofrequency transceiver. It is a low-cost device, easy to handle, and with sufficient signal reach.

The TRW24G model is based on the nRF2401 from Nordic with a completed circuit, which provides only the I/Os for the operation. Other characteristics of the transceiver are as follows: 1-Mbps transmission rate, multichannel operation (125 channels), 40-bit address, Cyclic Redundancy Check presence, simultaneous two-channel reception, 2.4-GHz operation, and 280 m of reach.

Modified Falcon Tube. A tube that allows gas exchange without spilling the culture medium during rotation was conceived and developed by the Microgravity Centre because there was no tube with the necessary characteristics for the experiment currently available.

A 50-mL modified Falcon tube was then used. A lid was placed on the Falcon tube, and three 6.5-mm holes were inserted to allow gas exchange between the cell culture and the environment. Between the lid of the tube and the tube itself, a membrane was also placed to avoid the culture medium spilling out. The whole set of lid, membrane, and tube is put into the sterilizer to prevent any kind of cell contamination.

Experimental Use of the Clinostat With Human Cells. The study protocol was approved by the Ethics Committee of the São Lucas Hospital. All subjects signed the informed consent form for the study and agreed with the protocol.

Subjects. Ten healthy subjects were selected to participate in the study (five men and five women; age, 24.7 ± 2.63 years). The exclusion criteria included any acute or chronic infection, heart and thyroid diseases, anemia, leukopenia, neoplasias, immunological depression, HIV infection or AIDS, and diabetes. None was taking any medication that would interfere with the analyses.

3.2 COLLECTION OF PERIPHERAL BLOOD AND ISOLATION OF MONONUCLEAR CELLS

Peripheral blood mononuclear cells (PBMCs) were used for the microgravity simulation tests. These cells (lymphocytes and monocytes) are part of the immunological system and are responsible for different immunological activities.

Twenty milliliters of peripheral blood was collected from the subjects by venopuncture in the morning (between 9 and 10 h), and the samples were then stored in lithium–heparin tubes before analyses. Samples were collected at the same time of day to minimize circadian variations. PBMCs were isolated by centrifugation over a Ficoll–Hypaque (Sigma®) gradient ($900g$, 30 min). Cells were counted by microscopy (100×) and viability always exceeded 95%, as evaluated by their ability to exclude trypan blue (Sigma).

After the samples were collected and isolated, they were divided into two tubes that were placed inside an incubator: one that was in clinorotation (24 and 48 h, 1.6 rpm) and a second one for the control at 1G (static). Neither of the samples was in contact with ultraviolet light or any kind of substance that could contaminate the cells.

3.2.1 Microgravity Test With PBMCs

After the isolation of the PBMCs, the cells were placed in the clinostat and exposed to 24 and 48 h of low-gravity environment (1.6 rpm) at 37°C in 5% CO_2 atmosphere. When the time (24 and 48 h) of microgravity simulation was completed, the cells were cultured in flat-bottomed 96-well microplates in a final concentration of 1.5×10^5 cells/well in complete culture medium (i.e., supplemented with gentamicin 0.5%, glutamine 1%, fungizone 0.1%, HEPES 1%, and heat-inactivated fetal calf serum 10%; all from Sigma) for 96 h in the same atmosphere as the 3D clinostat. Stimulation by the selective T-cell mitogen phytohemagglutinin (PHA 2, 1% and 0.5%; Gibco®, USA) was performed in triplicate (100 µL/well). In nonstimulated cultures (PHA 0), mitogen was substituted by culture medium.

3.3 CELL PROLIFERATION/VIABILITY ASSAY

The proliferative responses were determined by a modified colorimetric assay. In the last 4 h of culture, 100 µL of the supernatant was gently discarded and 40 µL of freshly prepared MTT

[3-(4,5-diamethyl 2-thiazolyl)2,5-diphenyl-2H-tetrazolium; Sigma) solution (5 mg/mL in RPMI-1640) was added to each well. The dehydrogenase enzymes in metabolically active cells convert this substrate to formazan, producing a dark blue precipitate. The cell cultures were incubated for 4 h at 37°C in 5% CO_2 atmosphere. After the complete removal of the supernatant, 100 μL of dimethyl sulfoxide (Sigma) was added to each well. The optical density (OD) was determined using Biorad enzyme-linked immunosorbent assay plate reader at a wavelength of 570 and 655 nm. Proliferation/viability was expressed as ΔOD (OD of stimulated − OD of nonstimulated cultures).

3.3.1   Statistical Analysis

Two-sample and paired tests were used for data analysis (95% confidence interval). A significance level of $P = .05$ was chosen.

3.3.2   Proliferation and Cellular Viability With PBMCs in Microgravity Simulation

The results are expressed in OD, which show the PBMC proliferation both in microgravity simulation and in static control. Table 3.2 presents the means ± SD of the cellular proliferation to the mitogen stimulation.

Results of the PBMCs tested during 24 and 48 h of microgravity simulation are also presented in Figure 3.21.

Figure 3.22 shows the results obtained during the proliferation and cell viability at 48 h of clinorotation.

TABLE 3.2: Mean cell proliferation to PHA stimulation during 24 and 48 h of clinorotation		
	24 H	48 H
PHA (%)	Mean ± SD	Mean ± SD
0.0	0.138 ± 0.071	0.076 ± 0.027
0.5	0.182 ± 0.091	0.134 ± 0.034
1.0	0.208 ± 0.138	0.171 ± 0.061
2.0	0.204 ± 0.196	0.144 ± 0.059

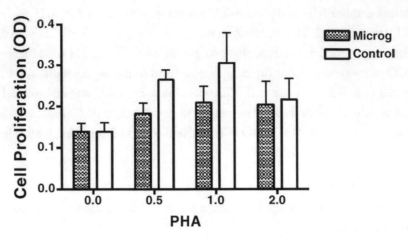

FIGURE 3.21: PBMC proliferation after 24 h of simulated microgravity (P = .146) (http://www.pucrs .br/feng/microg).

Figure 3.23 shows the comparison between the proliferation and cell viability at 24 and 48 h of microgravity simulation.

3.3.3 Proliferation and Viability Assay

The results indicate a nonsignificant decrease in the proliferation and cellular viability to the mitogen stimulation of PHA at 24 h of simulated weightlessness (P = .146) (Figure 3.21). These data

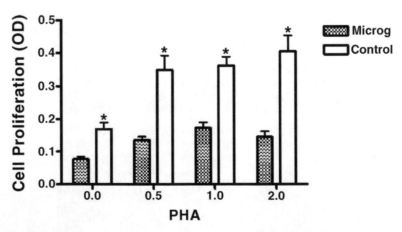

FIGURE 3.22: PBMC proliferation after 48 h of simulated microgravity (P = .012*) (http://www.pucrs .br/feng/microg).

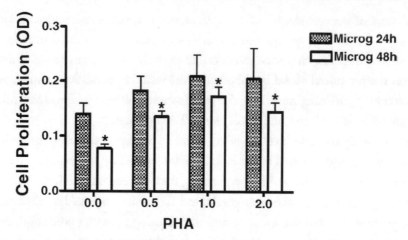

FIGURE 3.23: Comparison between cell culture in simulated microgravity during 24 and 48 h ($P = .003^*$) (http://www.pucrs.br/feng/microg).

indicate that for PBMCs, 24 h of clinorotation is not enough, which can be caused by the proliferation of mononuclear cells that occurs after 24 h. Walther et al (1998) showed that the expression of interleukin (IL)-2, a cytokine involved in proliferative response, is reduced by 85% after 46 h of clinorotation. It was then necessary to cultivate the mononuclear cells for 48 h of microgravity simulation, and the results confirmed a decrease in the proliferation and cellular viability ($P = .012$) (Figure 3.22).

Figure 3.23 represents a comparison between both microgravity simulation cultures at 24 and 48 h, and suggests that 48 h is the appropriate time to promote the expected effects of microgravity on mononuclear cells. Twenty-four hours of clinorotation may not be enough for that purpose. Therefore, it demonstrates that the proliferation of PBMCs decreases as the time of microgravity exposure increases. The next step will be to determine if this decrease in cell proliferation continues as the time of microgravity simulation increases.

The decrease of the proliferation of PBMCs may occur for different reasons. Hashemi et al (1999) found that the stimulation of PBMCs in clinorotation with PHA showed an inhibition of surface expression of CD25 (receptor of IL-2), CD69, and CD71, and the absence of CD25 expression in microgravity culture would certainly impair the responsiveness of T cells to the cytokine IL-2, an event that is required for the proliferative response. Clinorotation blocks T-cell transcription factor activation and inhibits the binding of transcription factors necessary to the production of IL-2 protein (Morrow, 2006).

Other studies indicate that the decrease observed in vitro are only related to gravitational effects occurring at the cellular level, whereas the ex vivo effects are caused by physical and psychological stress of the spaceflight on the immune system via neuroendocrine system (Taylor, 1993). Walther et al (1999) suggested that T-lymphocyte function is altered in more than 50% of space crewmembers, implying that such effect is due to stress rather than to weightlessness per se. However, when the peripheral blood lymphocytes were tested in conditions similar to real microgravity with random positioning machines, the results showed effects of simulated microgravity on the genetic expression of IL-2 and on its receptor in T lymphocytes.

The present study confirms that the effect of microgravity on the immune system is not just the physical and psychological stress of spaceflight but is caused by microgravity per se, which affects the proliferation and cellular viability, indicating that it occurs at cellular level.

Microgravity also affects other immunological functions, such as lymphocyte locomotion, which is an important aspect of the immune response. During growth, maturation, differentiation, and the inflammatory response, it is essential that lymphocytes traverse the extracellular matrix. Conditions that affect locomotion may have adverse consequences for human defense. Sundaresan et al (2002) had shown that lymphocyte locomotion was inhibited by 73% during 21 h of microgravity culture when compared with 1G cultured cells. A decreased production and function of certain types of cytokines, such as interferon γ, IL-1, and IL-2, are also observed in microgravity simulation (Sonnenfeld and Miller, 1993).

The present study indicates that immunological depression associated with spaceflight is not just related to the psychological and physiological stresses that the astronaut are subjected to, but it seems to be also caused by microgravity per se that affects the proliferation and cellular viability. The findings showed that the proliferative response to mitogen stimulation decrease in 48 h of clinorotation, which might impair the immune system of the person subjected to microgravity.

The clinostat was able to disorient the immunological cells and simulate microgravity on Earth. The results found in the study validate the Microgravity Centre 3D clinostat as a tool capable of simulating a weightlessness environment.

· · · ·

References

Agência Espacial Brasileira. Microgravidade. Available: http://www.inpe.br. accessed on November 22, 2007.

Alfrey CP, Udden MM, Laech-Huntoon C, Driscoll T, Picket MH. "Control of red blood cell mass in spaceflight". J Appl Physiol; 81:98–104, 1996.

Akima H, Katayama K, Sato K, Ishida K, Masuda K, Takada H, Watanabe Y, Iwase S. Intensive cycle training with artificial gravity maintains muscle size during bed rest. Aviat Space Environ Med; 76(10):923–929, Oct. 2005.

Audus LJ. "The Mechanism of the Perception of Gravity by Plants". In: Biologica Receptor Mechanism, No. XVI, Cambridge University Press, Cambridge, England. pp. 197–228, 1962.

Bungo MW, Charles JB, Johnson PC. "Cardiovascular deconditioning during spaceflight and the use of saline as a countermeasure to orthostatic intolerance". Aviat Space Environ Med; 56:985–990, 1985.

Brinckmann E. "New facilities and instruments for developmental biology research in space". Adv Space Biol Med; 9:53–80, 2003. doi:10.1016/S1569-2574(03)09010-5

Brinkley JW, Raddin JH Jr. "Biodynamic: Transitory Acceleration". In: Fundamentals of Aerospace Medicine. Chapter 8. DeHart RL (Ed.). Philadelphia, PA: Williams & Wilkins. pp. 162–201, 1985.

Bungo MW, Goldwater DJ, Popp RL, Sandler H. Echocardiographic evaluation of space shuttle crewmembers. J Appl Physiol; 62:278–283, 1987.

Burton RR. "The role of artificial gravity in the exploration of space". Acta Astronaut; 33: 217–220, Jul. 1994.

Burton RR, Whinnery JE. "Biodynamic: Sustained Acceleration". In: Fundamentals of Aerospace Medicine. Chapter 9. DeHart RL (Ed.). Philadelphia, PA: Williams & Wilkins. pp. 202–249, 1985.

Cardus D. "Artificial gravity in space and in medical research". J Gravit Physiol; 1(1):19–22, May 1994.

Chandler D. "Weightlessness and Microgravity". In: The Physics Teacher. Available: http://exploration.grc.nasa.gov/Exploration/redirect.htm. accessed on November 22, 2007. doi:10.1119/1.2343327

Coelho RP, Russomano T. Desenvolvimento de Centrífugas para Experimentos em Hipergravidade, M.S. thesis, Pontifical Catholic University, Porto Alegre, RS, Brazil, 2005.

Cutnell JD, and Johnson KW. In: Essentials of Physics. New Jersey: John Wiley and Sons, Inc. 2006.

Dobson K, Grace D, Lovett D, In: Physics. HaperCollins Publisher, Ltd., London, UK, 2006.

Draeger J, Schwartz R, Groenhoff S, Stern C. "2nd German D-2 Spacelab-Mission 1993: 114% IOP rise". Ophthalmologie; 91:697–699, 1994.

Ernsting J, Nicholson A, Rainford D. In: Aviation Medicine. 3rd ed. Oxford: Butterworth-Heinemann, 1999.

Edwards BF, Gray SW. "Chronic acceleration in plants". Life Sci Space Res; 15:273–278, 1977.

Forraz N, Russomano T, Falcão FP, Santos LGF, Motta JD, McGckin CP. "A novel microgravity-simulating bioreactor for tissue engineering". Paper presented at the 44th American Society for Cell Biology Annual Meeting, Washington, DC, December 5, 2004; 4–8, 2004.

Gell CF. "Table of Equivalents for Acceleration Terminology". Aviat Med; 32:1109–1111, 1961.

German Aerospace Centre. Available: http//www.dlr.de. accessed on November 22, 2007.

Iwase S. "Effectiveness of centrifuge-induced artificial gravity with ergometric exercise as a counter-measure during simulated microgravity exposure in humans". Acta Astronaut; 57(2–8):75–80, Jul.–Oct. 2005.

Halliday D, Resnick R, Walker J. "Vectors". In Fundamentals of Physics. Chapter 3. New York: John Wiley & Sons, Inc. pp. 97–130, 1993.

Hamalainen O, Toivakka-Hamalainen SK, Kuronen P. "+Gz associated stenosis of the cervical spinal canal in fighter pilots". Aviat Space Environ Med; 70(4):330–334, Apr. 1999.

Harding RM Mills FJ. In: Aviation Medicine. London: British Medical Association, 1983.

Hashemi BB, Penkala JE, Vens C, Huls H, Cubbage M, Sams CF. "T cell activation responses are differentially regulated during clinorotation and in spaceflight". FASEB J; 13:2071–2082, 1999.

Kakurin LI, Lobachik VI, Mikhailov VM, Senkevich YA. "Antiorthostatic Hypokinesia as a Method of Weightlessness Simulation". Aviat Space Environ Med; 47(10):1083–1086, 1976.

Klaus DM. "Clinostat and bioreactors". Grav Space Biol Bull; 14(2):55–64, 2001.

Lu SK, Bai S, Javeri K, Brunner LJ. "Altered cytochrome P450 and P-glycoprotein levels in rats during simulate weightlessness". Aviat Space Environ Med; 73:112–118, 2002.

LeBlanc AD, Evans HJ, Schneider VS, Wendt RE, Hedrick TD. "Changes in intervertebral disc cross-sectional area with bed rest and space flight". Spine; 19(7); 812–817, Apr. 1994.

Manti L. "Does reduced gravity alter cellular response to ionizing radiation?" Radiat Environ Biophys; 45:1–8, 2006.

Martinelli LK, Bauer ME, Russomano T. Aperfeiçoamento de um Clinostato 3d e Seu Uso no Estudo da Resistência a Múltiplas Drogas, M.S. thesis, Pontifical Catholic University, Porto Alegre, RS, Brazil, 2007. doi:10.1007/s00411-006-0037-4

Microgravity Centre/FENG-PUCRS. Available: http://www.pucrs.br/feng/microg/papers/2005_g .htm. accessed on November 22, 2007.

Microgravity Centre/FENG-PUCRS. Available: http://www.pucrs.br/feng/microg/papers/2006_o .htm. accessed on November 22, 2007.

Morrow MA. "Clinorotation differentially inhibits T-lymphocyte transcription factor activation". In Vitro Cell Dev Biol-Anim; 42:153–158, 2006.

Nagatomo, M. "Space tourism—spaceflight for the general public". J Practical Appl Space; 5:291–298, 1994.

Nemirovskaia T, Shenkman B. "Comparison of hypo- and hypergravity effects on skeletal muscle". J Gravit Physiol; 11(2);123–126, Jul. 2004.

Newman DJ. Life in extreme environments: how will humans perform on Mars? Gravit Space Biol Bull; 13(2):35–47, 2000.

Ohira YF, Goto K, Wang XD, Takeno Y, Ishihara A. "Role of gravity in mammalian development: effects of hypergravity and/or microgravity on the development of skeletal muscles". Biol Sci Space; 18(3): 124–125, Nov, 2004.

Russomano T, Chotgues LF, Sá OL, Santos M, Ernsting J. "Intra-ocular and venous pressures during head-down tilt". In: 49th International Astronautical Congress, 2001, Geneve. Scientific Program, v. 1. p. 26, 2001a.

Russomano, T, Azevedo DFG, Chotgues, LF. "Which angle of head-down tilt better simulates the increase in intra-ocular pressure that occurs in microgravity". In: 72nd Aerospace Medical Association Meeting, 2001, Reno, USA. 72nd AsMA Meeting—Book of Abstracts, v. 1. p. 73, 2001b.

Russomano T, Rizzatti MR, Coelho R, de Souza J, de Souza C, Falcão F, de Azevedo DFG, Scolari L. "Development of educational tools to demonstrate the effects of hyper and microgravity on different test bodies". Proceedings of the 26th Annual International Conference, IEEE Engineering in Medicine and Biology Society (EMBS), San Francisco, CA, USA, pp. 5141–5144, Sept. 2004.

Russomano T, Rizzatti MR, de Azevedo DFG, Coelho RP, Scolari D, Souza D, Prá-Veleda P. "Effects of simulated hypergravity on biomedical experiments". IEEE Eng Med Biol Mag, Inglaterra, Reino Unido; May/Jun:66–71. 2007.

Santos M, Forbes B, Marriott C. Evaluation of an intranasal scopolamine formulation for space motion sickness, Ph.D. dissertation, King's Colege London, University of London, London, UK, 2006. doi:10.1109/IEMBS.2004.1404431

Sastry KJ, Nehete PN, Savary CA. "Impairment of antigen-specific cellular immune responses under simulated microgravity conditions". In Vitro Cell Dev Biol-Anim; 37:203–208, 2001.

Smith AH. "Centrifuges: their development and use in gravitational biology". ASGSB Bull;5(2):33–41, Oct. 1992.

Sonnenfeld G, Miller ES. "The role of cytokines in immune changes induced by spaceflight". J Leukoc Biol; 54:253–258, 1993.

Strollo F. "Hormonal changes in humans during spaceflight". Adv Space Biol Med; 7:99–129, 1999.

Sundaresan A, Risin D, Pellis NR. "Loss of signal transduction and inhibition of lymphocyte locomotion in a ground based model microgravity". In Vitro Cell Biol.-Anim; 38:118–122, 2002.

Schwartz R, Draeger J, Groenhoff S, Flade KD. "1st German-Russian MIR mission: 92% IOP rise". Ophthalmologie; 90:640–642, 1993.

Taylor GR. "Overview of spaceflight immunology studies". J Leukoc Biol; 54:179–188, 1993. doi:10.1290/1071-2690(2002)038<0118:LOSTAI>2.0.CO;2

Unsworth BR, Lelkes PI. "Growing tissues in microgravity". Nat Med; 4:901–907, 1998.

Vieira A, Russomano T, Collin P, Falcão FP, Astarita L, Machado CA, dos Santos MA. "Estudo da germinação e do crescimento de *Eruca sativa* Mill. em simulação de hipergravidade". In: Salão de Iniciação Científica UFRGS, Porto Alegre, 2007.

Walther I, Pippia P, Meloni MA, Turrini F, Mannu F, Cogoli A. "Simulated microgravity inhibits the genetic expression of interleukin-2 and its receptor in mitogen-activated T lymphocytes". FEBS Lett; 436:115–118, 1998. doi:10.1038/nm0898-901

Walther I, Cogoli A, Pippia P, Meloni MA, Cossu G, Cogoli M, Schwarzenberg M, Turrini F, Mannu F. "Human immune cells as space travellers". Eur J Med Res; 4:361–363, 1999.

White WJ. A History of the Centrifuge in Aeroespace Medicine. Santa Monica: Douglas Aircraft Inc. p. 91, 1964. doi:10.1016/S0014-5793(98)01107-7

Young LR. "Artificial gravity considerations for a mars exploration mission". Ann N Y Acad Sci; 28:367–378, May 1999.

Author Biography

Thais Russomano graduated with a degree in medicine from the Federal University of Pelotas, Brazil, in 1985, a masters degree in aerospace medicine from Wright State University, United States, in 1991, and a doctor of philosophy degree in space physiology from King's College London, United Kingdom, in 1998.

She is the founder and coordinator of the internationally recognized Centre of Microgravity at PUCRS University Brazil (a unique reference center of Latin America in the study of human space physiology and space biomedical engineering), an associate professor of PUCRS University, a guest scientist at the German Space Agency, and a visiting professor/senior research fellow at King's College London.

She has more than 15 years experience in the fields of aerospace medicine, aerospace biomedicine, aerospace biomedical engineering and telemedicine, including participation in numerous scientific events and publication of more than 200 scientific articles. Research areas and professional experience include microgravity, hypogravity and hypergravity simulations, parabolic flights with the European Space Agency, hyperbaric and hypobaric chambers studies, tests in human centrifuges, rotator (Barany) chair, flight simulators, and lower body negative pressure boxes.

She won several national and international awards in the area of space life science research.

Felipe Prehn Falcão is an automation and control engineer and international researcher in aerospace biomedical engineering based at the Centre of Microgravity at PUCRS, Brazil. He has 5 years experience in the development of projects with research and teaching institutions in this area working with the German Space Agency, NASA, and King's College London. He has experience in aerospace medicine research including participation in a European Space Agency parabolic flight campaign (2006). He is involved in the development of hypergravity and microgravity simulation devices, lower-body negative-pressure box, rotator (Barany) chair, clinostat, human centrifuge, and biofeedback systems.

He was responsible for the conception and development of the Valsalva maneuver equipment, gaining national recognition and an award in the Siemens Prize of Technology Innovation for new projects.

He published many scientific articles and papers worldwide during congresses and via journals.

Gustavo Dalmarco is a biomedical engineer and international researcher in aerospace biomedical engineering. He developed projects with research and teaching institutions, such as the German Space Agency and King's College London and has five years experience working with aerospace medicine research including one European Space Agency parabolic flight campaign. He is involved in the development of hypergravity and microgravity simulation devices, lower body negative pressure box, human centrifuge, rotator (Barany) chairs, biofeedback systems, and clinostats.

Printed in the United States
by Baker & Taylor Publisher Services